SpringerBriefs in Meteorology

SpringerBriefs in Meteorology present concise summaries of cutting-edge research and practical applications. The series focuses on all aspects of meteorology including, but not exclusive to; tornadoes, thunderstorms, lightning, hail, rainfall, fog, extratropical and tropical cyclones, forecasting, snowfalls and blizzards, dust storms, clouds. The series also presents research and information on meteorological technologies, meteorological applications, meteorological forecasting and meteorological impacts (reports of notable worldwide weather events). Featuring compact volumes of 50–125 pages (approx. 20,000–70,000 words), the series covers a range of content from professional to academic such as: a timely reports of state-of-the art analytical techniques, literature reviews, in-depth case studies, bridges between new research results, snapshots of hot and/or emerging topics. Author Benefits: Books in this series will be published as part of Springer's eBook collection, with millions of users worldwide. In addition, Briefs will be available for individual print and electronic purchase. SpringerBriefs books are characterized by fast, global electronic dissemination and standard publishing contracts. Books in the program will benefit from easy-to-use manuscript preparation and formatting guidelines, and expedited production schedules. Both solicited and unsolicited manuscripts are considered for publication in this series. Projects will be submitted for editorial review by editorial advisory boards and/or publishing editors. For a proposal document please contact your Publisher, Dr. Robert K. Doe (robert.doe@springer.com).

More information about this series at http://www.springer.com/series/13553

A. M. Selvam

Rain Formation in Warm Clouds

General Systems Theory

Springer

A. M. Selvam
Pune
India

ISSN 2199-9112   ISSN 2199-9120 (electronic)
SpringerBriefs in Meteorology
ISBN 978-3-319-13268-6   ISBN 978-3-319-13269-3 (eBook)
DOI 10.1007/978-3-319-13269-3

Library of Congress Control Number: 2014959267

Springer Cham Heidelberg New York Dordrecht London

Printed on acid-free paper

Springer is part of Springer Science+Business Media (www.springer.com)

To my Parents: I. R. Muthusamy and Ida Muthusamy

Preface

Current concepts in meteorological theory and limitations are as follows. The nonequilibrium system of atmospheric flows is modeled with assumption of local thermodynamic equilibrium up to the stratopause at 50 km; molecular motion of atmospheric component gases is implicitly embodied in the gas constant. Nonequilibrium systems can be studied numerically, but despite decades of research, it is still very difficult to define the analytical functions from which to compute their statistics and have an intuition for how these systems behave. Realistic mathematical modeling for simulation and prediction of atmospheric flows requires alternative theoretical concepts and analytical or error-free numerical computational techniques and therefore comes under the field of "General Systems research" as explained in the following.

Space-time power law scaling and nonlocal connections exhibited by atmospheric flows have also been documented in other nonequilibrium dynamical systems, e.g. financial markets, neural network of brain, genetic networks, internet, road traffic, and flocking behavior of some animals and birds. Such universal behavior has been the subject of intensive study in recent years as "complex systems" under the subject headings self-organized criticality, nonlinear dynamics and chaos, network theory, pattern formation, information theory, cybernetics (communication, control, and adaptation). Complex system is a system composed of many interacting parts, such that the collective behavior or "emergent" behaviors of those parts together is more than the sum of their individual behaviors. Weather and climate are emergent properties of the complex adaptive system of atmospheric flows. Complex systems in different fields of study exhibit similar characteristics and therefore belong to the field of "General Systems." The terms "general systems" and "general systems research (or general systems theory)" are due to Ludwig von Bertalanffy. According to Bertalanffy, general systems research is a discipline whose subject matter is "the formulation and derivation of those principles which are valid for 'systems' in general."

Skyttner quotes basic ideas of general systems theory formulated by Fredrich Hegel (1770–1831) as follows:

1. The whole is more than the sum of the parts.
2. The whole defines the nature of the parts.

3. The parts cannot be understood by studying the whole.
4. The parts are dynamically interrelated or interdependent.

In cybernetics, a system is maintained in dynamic equilibrium by means of communication and control between the constituent parts and also between the system and its environment.

Chapter 1 gives applications of the concept of self-organized criticality to atmospheric flows. Atmospheric flows exhibit self-similar fractal fluctuations generic to dynamical systems in nature. Self-similarity implies long-range space-time correlations identified as self-organized criticality. It has been suggested that atmospheric convection could be an example of self–organized criticality. Atmospheric convection and precipitation have been hypothesized to be a real-world realization of self-organized criticality (SOC). The physics of self-organized criticality ubiquitous to dynamical systems in nature and in finite precision computer realizations of nonlinear numerical models of dynamical systems is not yet identified. Finite precision computer realizations of mathematical models (nonlinear) of dynamical systems do not give realistic solutions because of propagation of round off error into mainstream computation. During the past three decades, Lovejoy and his group have done extensive observational and theoretical studies of fractal nature of atmospheric flows and emphasize the urgent need to formulate and incorporate quantitative theoretical concepts of fractals in mainstream classical meteorological theory. The empirical analyses summarized by Lovejoy and Schertzer show that the statistical properties such as the mean and variance of atmospheric parameters (temperature, pressure, etc) are scale dependent and exhibit a power law relationship with a long fat tail over the space-time scales of measurement. The physics of the widely documented fractal fluctuations in dynamical systems is not yet identified. The traditional statistical normal (Gaussian) probability distribution is not applicable for statistical analysis of fractal space-time data sets because of the following reasons: (i) Gaussian distribution assumes independent (uncorrelated) data points while fractal fluctuations exhibit long-range correlations, (ii) the probability distribution of fractal fluctuations exhibit a long fat tail, i.e., extreme events are of more common occurrence than given by the classical theory.

A general systems theory model for fractal fluctuations proposed by the author predicts that the amplitude probability distribution as well as the power (variance) spectrum of fractal fluctuations follow the universal inverse power law $\tau^{-4\sigma}$ where τ is the golden mean (≈ 1.618) and σ the normalized standard deviation. The atmospheric aerosol size spectrum is derived in terms of the universal inverse power law characterizing atmospheric eddy energy spectrum. A universal (scale independent) spectrum is derived for homogeneous (same density) suspended atmospheric particulate size distribution expressed as a function of the golden mean τ (≈ 1.618), the total number concentration and the mean volume radius (or diameter) of the particulate size spectrum. Knowledge of the mean volume radius and total number concentration is sufficient to compute the total particulate size spectrum at any location. In summary, the model predictions are: (i) Fractal fluctuations can be resolved into an overall logarithmic spiral trajectory with the quasiperiodic Penrose tiling pattern for the internal structure. (ii) The probability distribution of fractal space-time fluctuations represents the power (variance) spectrum for fractal fluctuations and follows universal inverse power law form incorporating the golden mean. The result that the additive amplitudes of eddies when squared represent probability distribution is observed in the subatomic dynamics of quantum systems such as the electron or photon. Therefore, the irregular or unpredictable fractal fluctuations exhibit quantum-like chaos. (iii) Atmospheric aerosols are held in suspension by the vertical velocity distribution (spectrum). The normalized atmospheric aerosol size spectrum is derived in terms of the universal inverse power law characterizing atmospheric eddy energy spectrum. The model satisfies the maximum entropy principle.

In Chap. 2, the complete theory relating to the formation of warm cumulus clouds and their responses to the hygroscopic particle seeding are presented. It is shown that warm rain formation can occur within a time period of 30 min as observed in practice. Traditional cloud physical concepts for rain development requires over an hour for a full-sized raindrop to form.

Knowledge of the cloud dynamical, microphysical, and electrical parameters and their interactions are essential for the understanding of the formation of rain in warm clouds and their modification. Extensive aircraft observations of cloud dynamical, microphysical, and electrical parameters have been made in more than 2000 isolated warm cumulus clouds forming during the summer-monsoon seasons (June–September) in Pune (18° 32′N, 73° 51′E, 559 m asl), India. The observations were made during aircraft traverses at about 300 m above the cloud base. These observations have provided new evidence relating to the dynamics of monsoon clouds. The observed dynamical and physical characteristics of monsoon clouds cannot be explained by simple entraining cloud models. A simple cumulus cloud model which can explain the observed cloud characteristics has been developed. The relevant physical concept and theory relating to dynamics of atmospheric planetary boundary layer (PBL), formation of warm cumulus clouds, and their modification through hygroscopic particle seeding are presented.

The mechanism of large eddy growth discussed in the atmospheric ABL can be applied to the formulation of the governing equations for cumulus cloud growth. Based on the above theory equations are derived for the in-cloud vertical profiles of (i) ratio of actual cloud liquid water content (q) to the adiabatic liquid water content (qa) (ii) vertical velocity (iii) temperature excess (iv) temperature lapse rate (v) total liquid water content (qt) (vi) cloud growth time (vii) cloud drop size spectrum (viii) rain drop size spectrum. The equations are derived starting from the microscale fractional condensation (MFC) process at cloud base levels. This provides the basic energy input for the total cloud growth.

Chapter 3 discusses the importance of information on the size distribution of atmospheric suspended particulates (aerosols, cloud drops, raindrops) for the understanding of the physical processes relating to the studies in weather, climate, atmospheric electricity, air pollution, and aerosol physics. Atmospheric suspended particulates affect the radiative balance of the Earth/atmosphere system via the direct effect whereby they scatter and absorb solar and terrestrial radiation, and via the indirect effect whereby they modify the microphysical properties of clouds thereby affecting the radiative properties and lifetime of clouds. At present empirical models for the size distribution of atmospheric suspended particulates is used for the quantitative estimation of earth-atmosphere radiation budget related to climate warming/cooling trends. The empirical models for different locations at different atmospheric conditions, however, exhibit similarity in shape implying a common universal physical mechanism governing the organization of the shape of the size spectrum. The pioneering studies during the last three decades by Lovejoy and his group show that the particulates are held in suspension in turbulent atmospheric flows which exhibit self-similar fractal fluctuations on all scales ranging from turbulence (mm-s) to climate (km-years). Lovejoy and Schertzer have shown that the rain drop size distribution should show a universal scale invariant shape.

The general systems theory for fractal space-time fluctuations developed by the author predicts (Chap. 1) a universal (scale independent) spectrum for suspended atmospheric particulate size distribution expressed as a function of the golden mean τ (≈ 1.618), the total number concentration and the mean volume radius (or diameter) of the particulate size spectrum. Knowledge of the mean volume radius and total number concentration is sufficient to compute the total particulate size spectrum at any location. The model-predicted spectrum is in agreement with the following four experimentally determined data sets: (i) CIRPAS mission TARFOX_WALLOPS_ SMPS aerosol size distributions, (ii) CIRPAS mission ARM-IOP (Ponca City, OK) aerosol size distributions, (iii) SAFARI 2000 CV-580 (CARG Aerosol and Cloud Data) cloud drop size distributions, and (iv) TWP-ICE (Darwin, Australia) rain drop size distributions.

In Chap. 4 it is shown that the model-predicted suspended particulate (aerosol) size spectrum is in agreement with observations using VOCALS 2008 PCASP data.

In Chap. 5, the model-predicted spectrum is compared with the total averaged radius size spectra for the AERONET (aerosol inversions) stations Davos and Mauna Loa for the year 2010 and Izana for the year 2009.

In Chap. 6 it is shown that model-predicted spectrum is in agreement with the following two experimentally determined atmospheric aerosol data sets, (i) SAFARI 2000 CV-580 Aerosol Data, Dry Season 2000 (CARG) (ii) World Data Centre Aerosols data sets for the three stations Ny Ålesund, Pallas and Hohenpeissenberg.

There is close agreement between the model-predicted and the observed aerosol spectra at different locations. The proposed model for universal aerosol size spectrum will have applications in computations of radiation balance of earth–atmosphere system in climate models.

Indian Institute of Tropical Meteorology,   A. M. Selvam
Pune 411008, India   Deputy Director (Retired)
Address (Res): B1 Aradhana, 42/2A Shivajinagar,
Pune 411005, India
Email: amselvam@gmail.com; websites:
http://amselvam.webs.com; http://amselvam.tripod.com;
http://www.geocities.ws/amselvam

Contents

List of Frequently Used Symbols

v Frequency

d Aerosol diameter

N Aerosol number concentration

N^* Surface (or initial level) aerosol number concentration

r_a Aerosol radius

α Exponent of inverse power law

W Circulation speed (root mean square) of large eddy

w Circulation speed (root mean square) of turbulent eddy

R Radius of the large eddy

r Radius of the turbulent eddy

w^* Primary (initial stage) turbulent eddy circulation speed

r^* Primary (initial stage) turbulent eddy radius

T Time period of large eddy circulation

t Time period of turbulent eddy circulation

k Fractional volume dilution rate of large eddy by turbulent eddy fluctuations

z Eddy length scale ratio equal to R/r

f Steady state fractional upward mass flux of surface (or initial level) air

q Moisture content at height z

q^* Moisture content at primary (initial stage) level

m Suspended aerosol mass concentration at any level z

m^* Suspended aerosol mass concentration at primary (initial stage) level

r_a Mean volume radius of aerosols at level z

r_{as} Mean volume radius of aerosols at primary (initial stage) level

r_{an} Normalized mean volume radius equal to r_d/r_{as}

P Probability density distribution of fractal fluctuations

σ Normalized deviation

List of Frequently Used Symbols

Chapter 1
General Systems Theory Concepts in Atmospheric Flows

Abstract Atmospheric flows, a representative example of turbulent fluid flows, exhibit long-range spatiotemporal correlations manifested as the fractal geometry to the global cloud cover pattern concomitant with inverse power law form for spectra of temporal fluctuations. Such nonlocal connections are ubiquitous to dynamical systems in nature and are identified as signatures of self-organized criticality. Mathematical models for simulation and prediction of dynamical systems are nonlinear so that analytical solutions are not available. Finite precision computed solutions are sensitively dependent on initial conditions and give chaotic solutions, identified as deterministic chaos. Realistic mathematical modeling for simulation and prediction of atmospheric flows requires alternative theoretical concepts and analytical or error-free numerical computational techniques and therefore comes under the field of 'General Systems research'. General systems theory for atmospheric flows visualizes the hierarchical growth of larger scale eddies from space–time integration of smaller scale eddies resulting in an atmospheric eddy continuum manifested in the self-similar fractal fluctuations of meteorological parameters. The model shows that the observed long-range spatiotemporal correlations are intrinsic to quantum-like mechanics governing fluid flows. The model concepts are independent of intrinsic characteristics of the dynamical system such as chemical, physical, electrical, etc., and gives scale-free governing equations for fluid flow characteristics.

Keywords Nonlinear dynamics · Chaos · Fractals · Self-organized criticality · General systems theory

1.1 Introduction

Atmospheric flows exhibit self-similar fractal fluctuations generic to dynamical systems in nature. Self-similarity implies long-range space–time correlations identified as self-organized criticality (SOC) (Bak et al. 1988). Yano et al. (2012) suggest that atmospheric convection could be an example of self-organized criticality. Atmospheric convection and precipitation have been hypothesized to be a real-world realization of SOC (Peters et al. 2002, 2010). The physics of SOC ubiquitous to dynamical systems in nature and in finite precision computer realizations of nonlinear

© The Author(s) 2015   1
A. M. Selvam, *Rain Formation in Warm Clouds*, SpringerBriefs in Meteorology,
DOI 10.1007/978-3-319-13269-3_1

numerical models of dynamical systems is not yet identified. Finite precision computer realizations of mathematical models (nonlinear) of dynamical systems do not give realistic solutions because of propagation of round-off error into mainstream computation (Selvam 1993, 2007; Sivak et al. 2013, Lawrence Berkeley National Laboratory 2013). During the past three decades, Lovejoy and his group (Lovejoy and Schertzer 2010) have done extensive observational and theoretical studies of fractal nature of atmospheric flows and emphasize the urgent need to formulate and incorporate quantitative theoretical concepts of fractals in mainstream classical meteorological theory. The empirical analyses summarized by Lovejoy and Schertzer (2010) show that the statistical properties such as the mean and variance of atmospheric parameters (temperature, pressure, etc.) are scale dependent and exhibit a power law relationship with a long fat tail over the space–time scales of measurement. The physics of the widely documented fractal fluctuations in dynamical systems is not yet identified. The traditional statistical normal (Gaussian) probability distribution is not applicable for statistical analysis of fractal space–time data sets because of the following reasons: (i) Gaussian distribution assumes independent (uncorrelated) data points while fractal fluctuations exhibit long-range correlations; and (ii) the probability distribution of fractal fluctuations exhibit a long fat tail, i.e., extreme events are of more common occurrence than given by the classical theory (Selvam 2009; Lovejoy and Schertzer 2010).

Numerical computations such as addition, multiplication, etc., have inherent round-off errors. Iterative computations magnify these round-off errors because of feedback with amplification. Propagation of round-off errors into the main stream computation results in *deterministic chaos* in computer realizations of nonlinear mathematical models used for simulation of real-world dynamical systems. The round-off error growth structures generate the beautiful fractal patterns (Fig. 1.1). The complex structures found in nature are all *fractals*, i.e., possesses self-similar geometry. Simple iterative growth processes may underlie the complex patterns found in nature. The study of *fractals* belongs to the field of *nonlinear dynamics and chaos*, a multidisciplinary area of intensive research in all fields of science. Numerical integration schemes incorporate iterative computations.

Self-similar spatial structures imply long-range spatial correlations or nonlocal connections. Global cloud cover pattern exhibits *fractal* geometry. The existence of long-range spatial correlations such as the *El-Nino* impact on global climate is now accepted. The global atmosphere acts as a unified whole, where, local perturbations produce a global response.

A general systems theory model for fractal fluctuations (Selvam 2005, 2007, 1990, 2009, 2011, 2012a, b, 2013, 2014; Selvam and Fadnavis 1998) predicts that the amplitude probability distribution as well as the power (variance) spectrum of fractal fluctuations follow the universal inverse power law $\tau^{-4\sigma}$ where τ is the golden mean (≈ 1.618) and σ the normalized standard deviation. The atmospheric aerosol size spectrum is derived in terms of the universal inverse power law characterizing atmospheric eddy energy spectrum. A universal (scale independent) spectrum is derived for homogeneous (same density) suspended atmospheric particulate size distribution expressed as a function of the golden mean τ (≈ 1.618), the total number

Fig. 1.1 Graphic depiction of fractals generated using the Julia set algorithm, the simplest case for a chaotic attractor. The beautiful complex patterns of fractals generated by simple iterative computations consists of a hierarchy of self-similar structures, i.e., the large scale is a magnified version of the small scale

concentration and the mean volume radius (or diameter) of the particulate size spectrum. Knowledge of the mean volume radius and total number concentration is sufficient to compute the total particulate size spectrum at any location. In summary, the model predictions are (i) fractal fluctuations can be resolved into an overall logarithmic spiral trajectory with the quasiperiodic Penrose tiling pattern for the internal structure. (ii) The probability distribution of fractal space–time fluctuations represents the power (variance) spectrum for fractal fluctuations and follows universal inverse power law form incorporating the golden mean. Such a result that the additive amplitudes of eddies when squared represent probability distribution is observed in the subatomic dynamics of quantum systems such as the electron or photon. Therefore, the irregular or unpredictable fractal fluctuations exhibit quantum-like chaos. (iii) Atmospheric aerosols are held in suspension by the vertical velocity distribution (spectrum). The normalized atmospheric aerosol size spectrum is derived in terms of the universal inverse power law characterizing atmospheric eddy energy spectrum.

The complete theory relating to the formation of warm cumulus clouds and their responses to the hygroscopic particle seeding are presented. It is shown that warm rain formation can occur within a time period of 30 min as observed in practice. Traditional cloud physical concepts for rain development require over an hour for a full-sized raindrop to form (McGraw and Liu 2003). A review of droplet growth in warm clouds has been given by Devenish et al. (2012).

1.2 Current Concepts in Meteorological Theory and Limitations

The nonequilibrium system of atmospheric flows is modeled with assumption of local thermodynamic equilibrium up to the stratopause at 50 km; molecular motion of atmospheric component gases is implicitly embodied in the gas constant (Tuck 2010). Nonequilibrium systems can be studied numerically, but despite decades of research, it is still very difficult to define the analytical functions from which to compute their statistics and have an intuition for how these systems behave (Parmeggiani 2012). Realistic mathematical modeling for simulation and prediction of atmospheric flows requires alternative theoretical concepts and analytical or error-free numerical computational techniques and therefore comes under the field of "General Systems research" as explained in the following.

Space–time power law scaling and nonlocal connections exhibited by atmospheric flows have also been documented in other nonequilibrium dynamical systems, e.g., financial markets, neural network of brain, genetic networks, Internet, road traffic, flocking behavior of some animals and birds. Such universal behavior has been subject of intensive study in recent years as "complex systems" under the subject headings SOC, nonlinear dynamics and chaos, network theory, pattern formation, information theory, cybernetics (communication, control and adaptation). Complex system is a system composed of many interacting parts, such that the collective behavior or "emergent" behaviors of those parts together is more than the sum of their individual behaviors (Newman 2011). Weather and climate are emergent properties of the complex adaptive system of atmospheric flows. Complex systems in different fields of study exhibit similar characteristics and therefore belong to the field of "General Systems." The terms "general systems" and "general systems research (or general systems theory)" are due to Ludwig von Bertalanffy. According to Bertalanffy, general systems research is a discipline whose subject matter is "the formulation and derivation of those principles which are valid for 'systems' in general" (von Bertalanffy 1972; Klir 2001).

Skyttner (2005) quotes basic ideas of general systems theory formulated by Fredrich Hegel (1770–1831) as follows:

1. The whole is more than the sum of the parts
2. The whole defines the nature of the parts
3. The parts cannot be understood by studying the whole
4. The parts are dynamically interrelated or interdependent

In cybernetics, a system is maintained in dynamic equilibrium by means of communication and control between the constituent parts and also between the system and its environment (Skyttner 2005).

1.3   General Systems Theory for Fractal Space–Time Fluctuations in Atmospheric Flows

General systems theory for atmospheric flows is based on classical statistical physical concept where ensemble average represents the steady-state values of parameters such as pressure, temperature, etc., of molecular systems (gases) independent of details of individual molecule. The ideas of statistical mechanics (SM) have been successfully extended to various disciplines to study complex systems (Haken 1977; Liu and Daum 2001). Liu and his group (Liu 1992, 1995; Liu et al. 1995; Liu and Hallett 1997, 1998) have applied the systems approach to study cloud droplet size distributions.

Townsend (1956) had visualized large eddies as envelopes enclosing turbulent (smaller scale) eddies. General systems theory for atmospheric flows (Selvam 1990, 2005, 2012a, b, 2013; Selvam and Fadnavis 1998) visualizes the hierarchical growth of larger scale eddies from space–time integration of smaller scale eddies resulting in an atmospheric eddy continuum manifested in the self-similar fractal fluctuations of meteorological parameters. The basic thermodynamical parameters such as pressure, temperature, etc., are given by the same classical statistical physical formulae (kinetic theory of gases) for each component eddy (volume) of the atmospheric eddy continuum. It may be shown that the Boltzmann distribution for molecular energies also represents the eddy energy distribution in the atmospheric eddy continuum (Selvam 2011). In the following, general systems theory model concepts for atmospheric flows are summarized with model predictions for atmospheric flows and cloud growth parameters. Model predictions are compared with observations.

The atmospheric boundary layer (ABL), the layer extending up to about 10 km above the surface of the earth plays an important role in the formation of weather systems. It is important to identify and quantify the physical processes in the ABL for realistic simulation of weather systems of all scales.

The ABL is often organized into helical secondary circulations which are often referred to as vortex roll or large eddies (Brown 1980). It is not known how these vortex rolls are sustained without decay by the turbulence around them. The author (Selvam 1990, 2007) has shown that the production of buoyant energy by the microscale fractional condensation (MFC) in turbulent eddies is responsible for the sustenance and growth of large eddies. Earlier Eady (1950) has emphasized the importance of large-scale turbulence in the maintenance of the general circulation of the atmosphere.

The nondeterministic model described below incorporates the physics of the growth of macroscale coherent structures from microscopic domain fluctuations in atmospheric flows (Selvam 1990, 2007, 2013). In summary, the mean flow at the planetary ABL possesses an inherent upward momentum flux of frictional origin at the planetary surface. This turbulence-scale upward momentum flux is progressively amplified by the exponential decrease of the atmospheric density with height coupled with the buoyant energy supply by microscale fractional condensation on hygroscopic nuclei, even in an unsaturated environment (Pruppacher and Klett 1997). The mean large-scale upward momentum flux generates helical vortex-roll (or large eddy) circulations in the planetary ABL and is manifested as cloud rows and (or) streets, and mesoscale cloud clusters (MCC) in the global cloud cover pattern. A conceptual model (Selvam and Fadnavis 1998, Selvam 1990, 2005, 2007, 2009, 2011, 2012a, b, 2013) of large and turbulent eddies in the planetary ABL is shown in Figs. 1.2 and 1.3. The mean airflow at the planetary surface carries the signature of the fine scale features of the planetary surface topography as turbulent fluctuations with a net upward momentum flux. This persistent upward momentum flux of surface frictional origin generates large-eddy (or vortex-roll) circulations, which carry upward the turbulent eddies as internal circulations. Progressive upward growth of a

Fig. 1.2 Eddies in the atmospheric planetary boundary layer

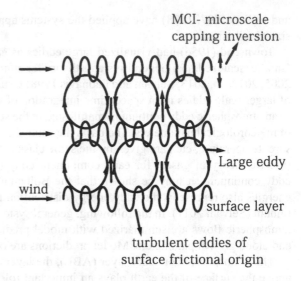

MCI- microscale
capping inversion

Large eddy

surface layer

wind

turbulent eddies of
surface frictional origin

Fig. 1.3 Growth of large eddy in the environment of the turbulent eddy

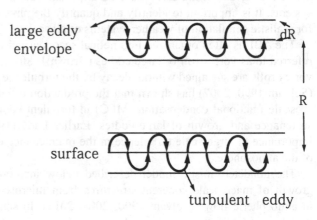

large eddy
envelope

dR

R

surface

turbulent   eddy

large eddy occurs because of buoyant energy generation in turbulent fluctuations as a result of the latent heat of condensation of atmospheric water vapor on suspended hygroscopic nuclei such as common salt particles. The latent heat of condensation generated by the turbulent eddies forms a distinct warm envelope or a microscale capping inversion (MCI) layer at the crest of the large-eddy circulations as shown in Fig. 1.2.

The turbulent eddies originating from surface friction exist all along the envelope of the large eddy (Fig. 1.2) and the MFC takes place even in an unsaturated environment (Pruppachar and Klett 1997).

Progressive upward growth of the large eddy occurs from the turbulence scale at the planetary surface to a height R and is seen as the rising inversion of the daytime ABL (Fig. 1.3).

The turbulent fluctuations at the crest of the growing large-eddy mix overlying environmental air into the large-eddy volume, i.e., there is a two-stream flow of warm air upward and cold air downward analogous to superfluid turbulence in liquid helium (Donelly 1998, 1990). The convective growth of a large eddy in the ABL therefore occurs by vigorous counter flow of air in turbulent fluctuations, which releases stored buoyant energy in the medium of propagation, e.g., latent heat of condensation of atmospheric water vapor. Such a picture of atmospheric convection is different from the traditional concept of atmospheric eddy growth by diffusion, i.e., analogous to the molecular level momentum transfer by collision (Selvam and Fadnavis 1998; Selvam 1990, 2005, 2007, 2009, 2011, 2012a, b, 2013).

The generation of turbulent buoyant energy by the microscale fractional condensation is maximum at the crest of the large eddies and results in the warming of the large-eddy volume. The turbulent eddies at the crest of the large eddies are identifiable by an MCI that rises upward with the convective growth of the large eddy during the course of the day. This is seen as the rising inversion of the daytime planetary boundary layer in echosonde and radiosonde records and has been identified as the entrainment zone (Boers 1989; Gryning and Batchvarova 2006), where mixing with the environment occurs.

The ABL contains large eddies (vortex rolls) which carry on their envelopes turbulent eddies of surface frictional origin (Selvam and Fadnavis 1998; Selvam 1990, 2005, 2007, 2009, 2011, 2012a, b, 2013). The buoyant energy production by *microscale-fractional condensation* (MFC) in turbulent eddies is responsible for the sustenance and growth of large eddies.

In summary, the buoyant energy production of turbulent eddies by the MFC process is maximum at the crest of the large eddies and results in the warming of the large eddy volume. The turbulent eddies at the crest of the large eddies are identifiable by a *microscale-capping inversion* (MCI) layer which rises upwards with the convective growth of the large eddy in the course of the day. The MCI layer is a region of enhanced aerosol concentrations. The atmosphere contains a stack of large eddies. Vertical mixing of overlying environmental air into the large eddy volume occurs by turbulent eddy fluctuations (Selvam and Fadnavis 1998; Selvam 1990, 2005, 2007, 2009, 2011, 2012a, b, 2013). The energy gained by the turbulent eddies would contribute to the sustenance and growth of the large eddy.

1.4   Large Eddy Growth in the ABL

Townsend (1956) has visualized the large eddy as the integrated mean of enclosed turbulent eddies. The root mean square (r.m.s) circulation speed W of the large eddy of radius R is expressed in terms of the enclosed turbulent eddy circulation speed w_* and radius r as (Selvam and Fadnavis 1998; Selvam 1990, 2005, 2007, 2009, 2011, 2012a, b, 2013).

$$W^2 = \frac{2}{\pi} \frac{r}{R} w_*^2. \tag{1.1}$$

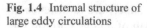

Fig. 1.4 Internal structure of
large eddy circulations

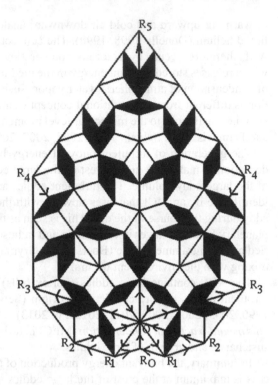

Based on Townsend's concept, the observed eddy continuum eddy growth in the
ABL is visualized to occur in the following two stages:

1. Growth of primary dominant turbulent eddy circulation from successive equal
   fluctuation length step increment dz equal to one. Identifiable organized whole
   turbulent eddy circulation forms at length step $z = 10$ associated with fractional
   volume dilution by eddy mixing less than half so that its (eddy) identity is not
   erased.
2. Large eddies are then visualized to form as envelopes enclosing these dominant
   turbulent eddies starting from unit primary eddy as zero level. Spatial domain
   of large eddy is obtained by integration from initial radius r to large eddy radius
   R (Eq. 1.1). The growing large eddy traces logarithmic spiral circulation with
   quasiperiodic Penrose tiling pattern (Fig. 1.4) for the internal structure such that
   successive large eddy radii follow the Fibonacci number series and ratio of successive radii R_{n+1}/R_n is equal to the golden mean τ (≈ 1.618).
3. Further it is shown that the variance (square of amplitude) spectrum and amplitude probability distribution of fractal fluctuations are represented by the same
   function, i.e., W and W^2 have the same probability distribution. Such a result that
   the additive amplitude of eddies when squared represent the probability distribution of amplitudes is exhibited by microscopic scale quantum systems such as
   the electron or photon. Fractal fluctuations therefore exhibit quantum-like chaos.
4. At each level, the integrated mean of eddy circulation speed w_* gives the
   mean circulation speed W for the next level. Therefore, W and w_* represent,

respectively, the mean and corresponding standard deviation of eddy circulation speed at that level and the ratio $W/w*$ is the normalized standard deviation σ. The primary turbulent eddy with circulation speed $w*$ and radius r is the reference level and normalized standard deviation σ values from 0 to 1 refer to the primary eddy growth region.

5. Primary eddy growth begins with unit length step perturbation followed by successive 10 unit length growth steps (Selvam 1990, 2012a, b, 2013; see Sect. 1.5 below).

1.5 Primary Dominant Eddy Growth Mechanism in the ABL

1.5.1 Steady-State Fractional Volume Dilution of Large Eddy by Turbulent Fluctuations

As seen from Figs. 1.2 and 1.3 and from the concept of large eddy growth, vigorous counter flow (mixing) in turbulent eddy fluctuations characterizes the large-eddy volume. The total fractional volume dilution rate of the large eddy by turbulent (eddy) vertical mixing of environmental air across unit cross-section of the large eddy surface is derived from Eq. (1.1) and is given as follows.

The ratio of the upward mass flux of air in the turbulent eddy to that in the large eddy across unit cross-section (of the large eddy) per second is equal to $w*/dW$, where $w*$ is the increase in vertical velocity per second of the turbulent eddy due to the MFC process, and dW is the corresponding increase in vertical velocity of large eddy. This fractional volume dilution of the large eddy occurs in the environment of the turbulent eddy. The fractional volume of the large eddy which is in the environment of the turbulent eddy where dilution occurs is equal to r/R.

Therefore, the total fractional volume dilution k of the large eddy per second across unit cross-section can be expressed as follows:

$$k = \frac{w_*}{dW} \frac{r}{R}. \tag{1.2}$$

The value of $k \approx 0.4$ when the length scale ratio R/r is equal to 10 since $dW \approx 0.25$ $w*$ (Eq. (1.1)). The growing large eddy cannot exist as a recognizable entity for length scale ratio values less than 10.

Identifiable large eddies can grow only for scale ratios $z > 10$. The convective scale eddy of radius R_c evolves from the turbulent eddy of radius r for the size ratio (z). $R_c/r = 10$. This type of decadic scale range eddy mixing can be visualized to occur in successive decadic scale ranges generating the convective, meso-, synoptic, and planetary scale eddies of radii R_c, R_m, R_s, and R_p where c, m, s, and p represent, respectively, the convective, meso-, synoptic, and planetary scales.

1.5.2   Logarithmic Wind Profile in the ABL

The height interval in which this incremental change dW in the vertical velocity occurs is dR which is equal to r. The height up to which the large eddy has grown is equal to R (see Fig. 1.3).

Using the above expressions Eq. (1.2) can be written as follows:

$$dW = \frac{w_*}{k}\frac{dR}{R} = \frac{w_*}{k}d\ln R. \qquad (1.3)$$

Integrating Eq. (1.3) between the height intervals r and R, the following relation for W can be obtained as follows:

$$W = \int_r^R \frac{w_*}{k}d\ln R = \frac{w_*}{k}(\ln R - \ln r)$$

$$W = \frac{w_*}{k}\ln\left(\frac{R}{r}\right) = \frac{w_*}{k}\ln z. \qquad (1.4)$$

In Eq. (1.4), it is assumed that w_* and r_* are constant for the height interval of integration. The length scale ratio R/r is denoted by z. A normalized height with reference to the turbulence scale r can be defined as $z = R/r$.

Equation (1.4) is the well-known logarithmic velocity profiles in turbulent shear flows discussed originally by von Karman (1956) and Prandtl (1932) (for a recent review, see Marusic et al. 2010), and to the recently discovered logarithmic variation of turbulent fluctuations in pipe flow (Hultmark et al. 2012). Observations and the existing theory of eddy diffusion (Holton 2004) indicate that the vertical wind profile in the ABL follows the logarithmic law which is identical to the expression shown in Eq. (1.4). The constant k (Von Karman's constant) as determined from observations is equal to 0.4 and has not been assigned any physical meaning in the literature. The new theory proposed in this study enables prediction of observed logarithmic wind profile without involving any assumptions as in the case of existing theories of atmospheric diffusion processes such as molecular momentum transfer (Holton 2004). Also, the constant k now has a physical meaning, namely, it is the fractional volume dilution rate of the large eddy by the turbulent scale eddies for dominant large eddy growth.

1.5.3   Fractional Upward Mass Flux of Surface Air

Vertical mixing due to turbulent eddy fluctuations progressively dilutes the rising large eddy and a fraction f equal to $Wr/w_* R$ of surface air reaches the normalized height z as shown in the following. The turbulent eddy fluctuations carry upward surface air of frictional origin. The ratio of upward mass flux of air/unit time/unit cross-section in the large eddy to that in the turbulent eddy $= W/w_*$. The magnitude of W is smaller than that of w_* (Eq. 1.1). The ratio W/w_* is equal to the upward mass flux of surface air/unit time/unit cross-section in the rising large eddy. The volume fraction of turbulent eddy across unit cross-section of large eddy envelope is equal to r/R. Turbulence scale upward mass flux of surface air equal to W/w_* occurs in the fractional volume r/R of the large eddy. Therefore, the net upward mass flux f of surface air/unit time/unit cross-section in the large eddy environment is equal to

$$f = \frac{W}{w_*} \frac{r}{R}. \tag{1.5}$$

The large eddy circulation speed W and the corresponding temperature perturbation θ may be expressed in terms of f and z as follows:

$$W = w_* fz$$
$$\theta = \theta_* fz. \tag{1.6}$$

In Eq. (1.6), θ_* is the temperature perturbation corresponding to the primary turbulent eddy circulation speed w_*.

The corresponding moisture content q at height z is related to the moisture content q_* at the surface and is given as follows:

$$q = q_* fz. \tag{1.7}$$

Substituting from Eqs. (1.1), (1.2), and (1.4), the net upward flux f of surface air at level z is now obtained from Eq. (1.5) as follows:

$$f = \frac{1}{kz} \ln z = \frac{WR}{w_* rz} \ln z = \sqrt{\frac{2}{\pi z}} \ln z. \tag{1.8}$$

In Eq. (1.8), f represents the steady-state fractional volume of surface air at any level z. A fraction f of surface aerosol concentration N_* is carried upward to normalized height z. The aerosol number concentration N at level z is then given as follows:

$$N = N_* f. \tag{1.9}$$

Since atmospheric aerosols originate from surface, the vertical profile of mass and number concentration of aerosols follow the f distribution. The vertical mass exchange mechanism predicts the f distribution for the steady-state vertical transport of aerosols at higher levels. The vertical variation of atmospheric aerosol number concentration given by the f distribution is shown in Fig. 1.5. The vertical variation of large eddy circulation speed W, the corresponding temperature θ and the moisture content q are shown in Fig. 1.6.

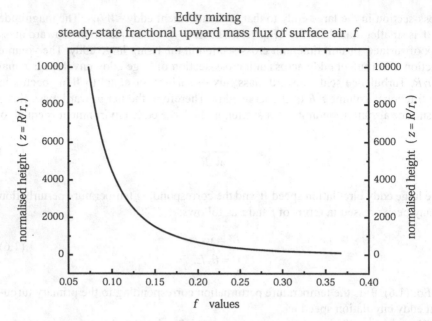

Fig. 1.5 f distribution for $r^* = 1$. The above f distribution represents vertical distribution of (i) atmospheric aerosol concentration (ii) ratio of cloud liquid water content to adiabatic liquid water content

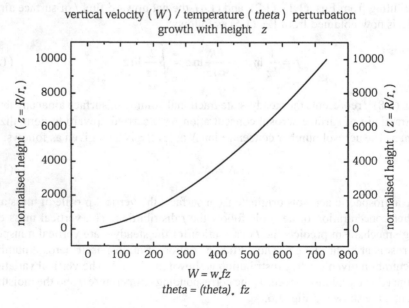

Fig. 1.6 Vertical variation of large eddy circulation speed W, temperature θ, and moisture content q

1.6 Atmospheric Aerosol (Particulates) Size Spectrum

The atmospheric eddies hold in suspension the aerosols and thus the size spectrum of the atmospheric aerosols is dependent on the vertical velocity spectrum of the atmospheric eddies as shown below. Earlier Liu (1956) has studied the problem of the dispersion of material particles in a turbulent fluid and remarks that particle dispersion constitutes a direct and striking manifestation of the mechanism of fluid turbulence. Grabowskii and Wang (2013) discuss multiscale nature of turbulent cloud microphysical processes and its significant impact on warm rain initiation.

The aerosols are held in suspension by the eddy vertical velocity perturbations. Thus, the suspended aerosol mass concentration m at any level z will be directly related to the vertical velocity perturbation W at z, i.e., $W \sim mg$, where g is the acceleration due to gravity. Substituting in Eq. (1.6) for W and $w*$ in terms of aerosol mass concentrations m and $m*$, respectively, at normalized height z and at surface layer, the vertical variation of aerosol mass concentration flux is obtained as follows:

$$m = m* fz. \tag{1.10}$$

1.6.1 Vertical Variation of Aerosol Mean Volume Radius

The mean volume radius of aerosol increases with height as shown in the following.

The velocity perturbation W is represented by an eddy continuum of corresponding size (length) scales z. The aerosol mass flux across unit cross-section per unit time is obtained by normalizing the velocity perturbation W with respect to the corresponding length scale z to give the volume flux of air equal to Wz and can be expressed as follows from Eq. (1.6):

$$Wz = (w* fz)z = w* fz^2. \tag{1.11}$$

The corresponding normalized moisture flux perturbation is equal to qz, where q is the moisture content per unit volume at level z. Substituting for q from Eq. (1.7)

$$qz = normalized\ moisture\ flux\ at\ level\ z = q* fz^2. \tag{1.12}$$

The moisture flux increases with height resulting in increase of mean volume radius of cloud condensation nuclei (CCN) because of condensation of water vapor. The corresponding CCN (aerosol) mean volume radius r_a at height z is given in terms of the aerosol number concentration N at level z and mean volume radius r_{as} at the surface as follows from Eq. (1.12)

$$\frac{4}{3}\pi r_a^3 N = \frac{4}{3}\pi r_{as}^3 N* fz^2. \tag{1.13}$$

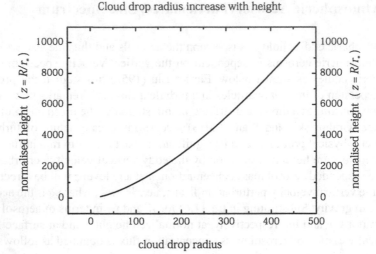

Fig. 1.7 Vertical variation of aerosol mean volume radius with height

Substituting for N from Eq. (1.9) in terms of N_* and f

$$r_a^3 = r_{as}^3 z^2$$
$$r_a = r_{as} z^{2/3} .$$
(1.14)

The mean aerosol size increases with height according to the cube root of z^2 (Eq. 1.14). Vertical variation of aerosol mean volume radius with height is shown in Fig. 1.7. As the large eddy grows in the vertical, the aerosol size spectrum extends towards larger sizes while the total number concentration decreases with height according to the f distribution.

The atmospheric aerosol size spectrum is dependent on the eddy energy spectrum and may be expressed in terms of the recently identified universal characteristics of fractal fluctuations generic to atmospheric flows (Selvam and Fadnavis 1998; Selvam 1990, 2005, 2007, 2009, 2011, 2012a, b, 2013) as shown in Sect. 1.6.2 below.

1.6.2 Probability Distribution of Fractal Fluctuations in Atmospheric Flows

The atmospheric eddies hold in suspension atmospheric particulates, namely, aerosols, cloud drops and raindrops and the size spectrum of these atmospheric suspended particulates is dependent on the vertical velocity spectrum of the atmospheric eddies. Atmospheric air flow is turbulent, i.e., consists of irregular fluctuations of all space–time scales characterized by a broadband spectrum of eddies. The suspended particulates will also exhibit a broadband size spectrum closely related to the atmospheric eddy energy spectrum.

It is now established (Lovejoy and Schertzer 2010) that atmospheric flows exhibit self-similar fractal fluctuations generic to dynamical systems in nature such as fluid flows, heart-beat patterns, population dynamics, spread of forest fires, etc. Power spectra of fractal fluctuations exhibit inverse power law of form $-f^{\alpha}$, where α is a constant indicating long-range space–time correlations or persistence. Inverse power law for power spectrum indicates scale invariance, i.e., the eddy energies at two different scales (space–time) are related to each other by a scale factor (α in this case) alone independent of the intrinsic properties such as physical, chemical, electrical, etc., of the dynamical system.

A general systems theory for turbulent fluid flows predicts that the eddy energy spectrum, i.e., the variance (square of eddy amplitude) spectrum is the same as the probability distribution P of the eddy amplitudes, i.e., the vertical velocity W values. Such a result that the additive amplitudes of eddies, when squared, represent the probabilities is exhibited by the subatomic dynamics of quantum systems such as the electron or photon. Therefore, the unpredictable or irregular fractal space–time fluctuations generic to dynamical systems in nature, such as atmospheric flows is a signature of quantum-like chaos. The general systems theory for turbulent fluid flows predicts (Selvam and Fadnavis 1998; Selvam 1990, 2005, 2007, 2009, 2011, 2012a, b, 2013) that the atmospheric eddy energy spectrum represented by the probability distribution P follows inverse power law form incorporating the *golden mean* τ and the normalized deviation σ for values of $\sigma \geq 1$ and $\sigma \leq -1$ as given below:

$$P = \tau^{-4\sigma}. \tag{1.15}$$

The vertical velocity W spectrum will, therefore, be represented by the probability distribution P for values of $\sigma \geq 1$ and $\sigma \leq -1$ given in Eq. (1.15) since fractal fluctuations exhibit quantum-like chaos as explained above.

$$W = P = \tau^{-4\sigma}. \tag{1.16}$$

Values of the normalized deviation σ in the range $-1 < \sigma < 1$ refer to regions of primary eddy growth where the fractional volume dilution k (Eq. 1.2) by eddy mixing process has to be taken into account for determining the probability distribution P of fractal fluctuations (see Sect. 1.6.3 below).

1.6.3   Primary Eddy Growth Region Fractal Fluctuation Probability Distribution

Normalized deviation σ ranging from -1 to $+1$ corresponds to the primary eddy growth region. In this region, the probability P is shown to be equal to $P = \tau^{-4k}$ (see below) where k is the fractional volume dilution by eddy mixing (Eq. 1.2).

The normalized deviation σ represents the length step growth number for growth stages more than one. The first stage of eddy growth is the primary eddy growth starting from unit length scale ($r = 1$) perturbation, the complete eddy forming at the tenth length scale growth, i.e., $R = 10 \, r$ and scale ratio z equals 10 (Selvam 1990, 2012a, b, 2013). The steady-state fractional volume dilution k of the growing primary eddy by internal smaller scale eddy mixing is given by Eq. (1.2) as follows:

$$k = \frac{w_* r}{WR}. \tag{1.17}$$

The expression for k in terms of the length scale ratio z equal to R/r is obtained from Eq. (1.1) as

$$k = \sqrt{\frac{\pi}{2z}}. \tag{1.18}$$

A fully formed large eddy length $R = 10r$ ($z = 10$) represents the average or mean level zero and corresponds to a maximum of 50% cumulative probability of occurrence of either positive or negative fluctuation peak at normalized deviation σ value equal to 0 by convention. For intermediate eddy growth stages, i.e., z less than 10, the probability of occurrence of the primary eddy fluctuation does not follow conventional statistics, but is computed as follows taking into consideration the fractional volume dilution of the primary eddy by internal turbulent eddy fluctuations. Starting from unit length scale fluctuation, the large eddy formation is completed after 10 unit length step growths, i.e., a total of 11 length steps including the initial unit perturbation. At the second step ($z = 2$) of eddy growth the value of normalized deviation σ is equal to 1.1−0.2 (=0.9) since the complete primary eddy length plus the first length step is equal to 1.1. The probability of occurrence of the primary eddy perturbation at this σ value however, is determined by the fractional volume dilution k which quantifies the departure of the primary eddy from its undiluted average condition and therefore represents the normalized deviation σ. Therefore, the probability density P of fractal fluctuations of the primary eddy is given using the computed value of k as shown in the following equation:

$$P = \tau^{-4k}. \tag{1.19}$$

The vertical velocity W spectrum will, therefore, be represented by the probability density distribution P for values of $-1 \leq \sigma \leq 1$ given in Eq. (1.19) since fractal fluctuations exhibit quantum-like chaos as explained above (Eq. 1.16).

$$W = P = \tau^{-4k}. \tag{1.20}$$

The probabilities of occurrence (P) of the primary eddy for a complete eddy cycle either in the positive or negative direction starting from the peak value (σ=0) are given for progressive growth stages (σ values) in the following Table 1.1. The statis-

Table 1.1 Primary eddy growth

Growth step number	±σ	k	Probability (%)	
			Model predicted	Statistical normal
2	0.9000	0.8864	18.1555	18.4060
3	0.8000	0.7237	24.8304	21.1855
4	0.7000	0.6268	29.9254	24.1964
5	0.6000	0.5606	33.9904	27.4253
6	0.5000	0.5118	37.3412	30.8538
7	0.4000	0.4738	40.1720	34.4578
8	0.3000	0.4432	42.6093	38.2089
9	0.2000	0.4179	44.7397	42.0740
10	0.1000	0.3964	46.6250	46.0172
11	0	0.3780	48.3104	50.0000

tical normal probability density distribution corresponding to the normalized deviation σ values are also given in Table 1.1.

The model predicted probability density distribution P along with the corresponding statistical normal distribution with probability values plotted on linear and logarithmic scales, respectively, on the left- and right-hand sides are shown in Fig. 1.8. The model predicted probability distribution P for fractal space–time fluctuations is very close to the statistical normal distribution for normalized deviation σ values less than 2 as seen on the left-hand side of Fig. 1.8. The model predicts progressively higher values of probability P for values of σ greater than 2 as seen on a logarithmic plot on the right-hand side of Fig. 1.8.

1.6.4 Atmospheric Wind Spectrum and Suspended Particulate Size Spectrum

The steady-state flux dN of CCN at level z in the normalized vertical velocity perturbation $(dW)z$ is given as follows:

$$dN = N(dW)z. \tag{1.21}$$

The logarithmic wind profile relationship for W at Eq. (1.4) gives

$$dN = Nz\frac{w_*}{k}d(\ln z). \tag{1.22}$$

Substituting for k from Eq. (1.2)

$$dN = Nz\frac{w_*}{w_*}Wzd(\ln z) = NWz^2d(\ln z). \tag{1.23}$$

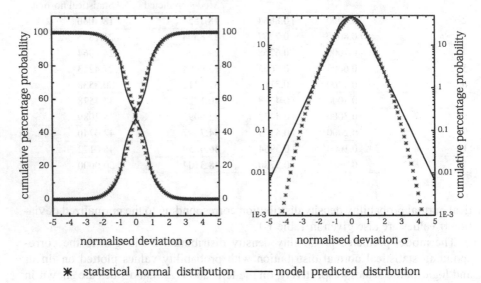

fractal fluctuations probability distribution comparison with statistical normal distribution

✳ statistical normal distribution   ——— model predicted distribution

Fig. 1.8 Model predicted probability distribution P along with the corresponding statistical normal distribution with probability values plotted on linear and logarithmic scales, respectively, on the *left*-and *right*-hand side graphs

The length scale z is related to the suspended particulate radius r_a (Eq. 1.14). Therefore,

$$\ln z = \frac{3}{2} ln \left(\frac{r_a}{r_{as}} \right). \tag{1.24}$$

Defining a normalized radius r_{an} equal to $\frac{r_a}{r_{as}}$, i.e., r_{an} represents the suspended particulate mean volume radius r_a in terms of its mean volume radius r_{as} at the surface (or reference level). Therefore,

$$\ln z = \frac{3}{2} \ln r_{an}, \tag{1.25}$$

$$d \ln z = \frac{3}{2} d \ln r_{an}. \tag{1.26}$$

Substituting for dln$z$ in Eq. (1.23)

$$dN = NWz^2 \frac{3}{2} d(\ln r_{an}), \tag{1.27}$$

$$\frac{dN}{d(\ln r_{an})} = \frac{3}{2} NWz^2. \tag{1.28}$$

Substituting for W from Eqs. (1.16) and (1.20) in terms of the universal probability density P for fractal fluctuations

$$\frac{dN}{d(\ln r_{an})} = \frac{3}{2} NPz^2. \tag{1.29}$$

The above equation is for the scale length z. The volume across unit cross-section associated with scale length z is equal to z. The particle radius corresponding to this volume is equal to $z^{1/3}$.

The above Eq. (1.28) is for the scale length z and the corresponding radius equal to $z^{1/3}$. The Eq. (1.28) normalized for scale length and associated drop radius is given as follows:

$$\frac{dN}{d(\ln r_{an})} = \frac{3}{2} \frac{NPz^2}{z \times z^{\frac{1}{3}}} = \frac{3}{2} NPz^{\frac{2}{3}}. \tag{1.30}$$

The general systems theory predicts that fractal fluctuations may be resolved into an overall logarithmic spiral trajectory with the quasiperiodic Penrose tiling pattern for the internal structure such that the successive eddy lengths follow the Fibonacci mathematical number series (Selvam and Fadnavis 1998; Selvam 1990, 2005, 2007, 2009, 2011, 2012a, b, 2013). The eddy length scale ratio z for length step σ is, therefore, a function of the golden mean τ given as follows:

$$z = \tau^{\sigma}. \tag{1.31}$$

Expressing the scale length z in terms of the golden mean τ in Eq. (1.29)

$$\frac{dN}{d(\ln r_{an})} = \frac{3}{2} NP\tau^{\frac{2\sigma}{3}}. \tag{1.32}$$

In Eq. (1.32), N is the steady-state aerosol concentration at level z. The normalized aerosol concentration any level z is given as follows:

$$\frac{1}{N} \frac{dN}{d(\ln r_{an})} = \frac{3}{2} P\tau^{\frac{2\sigma}{3}}. \tag{1.33}$$

The fractal fluctuations probability density is $P = \tau^{-4\sigma}$ (Eq. 1.16) for values of the normalized deviation $\sigma \geq 1$ and $\sigma \leq -1$ on either side of $\sigma = 0$ as explained earlier (Sects. 1.6.2 and 1.6.3). Values of the normalized deviation $-1 \leq \sigma \leq 1$ refer to regions of primary eddy growth where the fractional volume dilution k (Eq. 1.2) by eddy mixing process has to be taken into account for determining the probability density P of fractal fluctuations. Therefore, the probability density P in the primary eddy growth region ($\sigma \geq 1$ and $\sigma \leq -1$) is given using the computed value of k as $P = \tau^{-4k}$ (Eq. 1.20).

The normalized radius r_{an} is given in terms of σ and the golden mean τ from Eqs. (1.25) and (1.31) as follows:

$$\ln z = \frac{3}{2} \ln r_{an} \qquad\qquad (1.34)$$

$$r_{an} = z^{2/3} = \tau^{2\sigma/3}.$$

The normalized aerosol size spectrum is obtained by plotting a graph of normalized aerosol concentration $\dfrac{1}{N}\dfrac{dN}{d(\ln r_{an})} = \dfrac{3}{2} P \tau^{\frac{2\sigma}{3}}$ (Eq. 1.33) versus the normalized aerosol radius $r_{an} = \tau^{2\sigma/3}$ (Eq. 1.34). The normalized aerosol size spectrum is derived directly from the universal probability density P distribution characteristics of fractal fluctuations (Eqs. 1.16 and 1.20) and is independent of the height z of measurement and is universal for aerosols in turbulent atmospheric flows. The aerosol size spectrum is computed starting from the minimum size, the corresponding probability density P (Eq. 1.33) refers to the cumulative probability density starting from 1 and is computed as equal to $P = 1 - \tau^{-4\sigma}$.

The universal normalized aerosol size spectrum represented by $\dfrac{1}{N}\dfrac{dN}{d(\ln r_{an})}$ versus r_{an} is shown in Fig. 1.9.

1.6.5   Large Eddy Growth Time

The time Γ taken for the steady-state aerosol concentration f to be established at the normalized height z is equal to the time taken for the large eddy to grow to the height z and is computed as follows.

The time required for the large eddy of radius R to grow from the primary turbulence scale radius r_* is computed as follows.

The scale ratio $z = \dfrac{R}{r_*}$.

Therefore, for constant turbulence radius r_*

$$dz = \frac{dR}{r_*}.$$

The incremental growth dR of large eddy radius is equal to

$$dR = r * dz.$$

The time period dt for the incremental cloud growth is expressed as follows:

Model predicted aerosol size spectrum

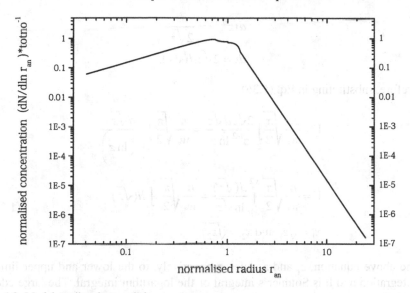

Fig. 1.9 Model predicted aerosol size spectrum

$$dt = \frac{dR}{W} = \frac{r_* dz}{W}. \tag{1.35}$$

The large eddy circulation speed W is expressed in terms of f and z as (Eq. 1.6)

$$W = w_* fz.$$

Substituting for f from Eq. (1.8)

$$W = w_* z \sqrt{\frac{2}{\pi z}} \ln z = w_* \sqrt{\frac{2z}{\pi}} \ln z.$$

Substituting for W in Eq. (1.35), the incremental eddy growth time dt is given as follows:

$$dt = \frac{r_* dz}{w_* fz} = \frac{r_* dz}{w_* z \sqrt{\dfrac{2}{\pi z}} \ln z}.$$

The time Γ taken for large eddy growth from surface to normalized height z is obtained as follows:

$$\Gamma = \int dt = \frac{r_*}{w_*} \sqrt{\frac{\pi}{2}} \int_2^z \frac{dz}{z^{1/2} \ln z}. \tag{1.36}$$

The above equation can be written in terms of $\sqrt{z}$ as follows:

$$d(z^{0.5}) = \frac{dz}{2\sqrt{z}}$$

$$dz = 2\sqrt{z}d(\sqrt{z}).$$

Therefore, substituting in Eq. (1.36)

$$\Gamma = \frac{r_*}{w_*}\sqrt{\frac{\pi}{2}}\int_2^z \frac{2\sqrt{z}d\sqrt{z}}{z^{1/2}\ln z} = \frac{r_*}{w_*}\sqrt{\frac{\pi}{2}}\int_2^z \frac{d\sqrt{z}}{\left(\frac{1}{2}\ln z\right)}$$

$$\Gamma = \frac{r_*}{w_*}\sqrt{\frac{\pi}{2}}\int_{x1}^{x2} \frac{d(\sqrt{z})}{\ln\sqrt{z}} = \frac{r_*}{w_*}\sqrt{\frac{\pi}{2}}\int_{x1}^{x2} li(\sqrt{z}) \qquad (1.37)$$

$$x_1 = \sqrt{z_1} \text{ and } x_2 = \sqrt{z_2}.$$

In the above equation, z_1 and z_2 refer, respectively, to the lower and upper limits of integration and li is Soldner's integral or the logarithm integral. The large eddy growth time Γ can be computed from Eq. (1.37).

1.7   General Systems Theory and Classical Statistical Physics

A summary of Lebowitz's (Lebowitz 1999) discussion on the essential role of classical statistical mechanical concepts underlying the formulation of precise physical laws for observed macroscale phenomena in nature is given below. Nature has a hierarchical structure, with time, length, and energy scales ranging from the submicroscopic to the supergalactic. Surprisingly it is possible and in many cases essential to discuss these levels independently—quarks are irrelevant for understanding protein folding and atoms are a distraction when studying ocean currents. Nevertheless, it is a central lesson of science, very successful in the past 300 years, that there are no new fundamental laws, only new phenomena, as one goes up the hierarchy. Thus, arrows of explanations between different levels always point from smaller to larger scales, although the origin of higher level phenomena in the more fundamental lower level laws is often very far from transparent. SM provides a framework for describing how well-defined higher level patterns or behavior may result from the nondirected activity of a multitude of interacting lower level individual entities. The subject was developed for, and has had its greatest success so far in, relating mesoscopic and macroscopic thermal phenomena to the microscopic world of atoms and molecules. SM explains how macroscopic phenomena originate in the cooperative behavior of these microscopic particles (Lebowitz 1999).

The general systems theory visualizes the self-similar fractal fluctuations to result from a hierarchy of eddies, the larger scale being the space–time average of enclosed smaller scale eddies (Eq. 1.1) assuming constant values for the characteristic length scale r and circulation speed $w*$ throughout the large eddy space–time domain. The collective behavior of the ordered hierarchical eddy ensembles is manifested as the apparently irregular fractal fluctuations with long-range space–time correlations generic to dynamical systems. The concept that aggregate averaged eddy ensemble properties represent the eddy continuum belongs to the nineteenth-century classical statistical physics where the study of the properties of a system is reduced to a determination of average values of the physical quantities that characterize the state of the system as a whole (Yavorsky and Detlaf 1975) such as gases, e.g., the gaseous envelope of the earth, the atmosphere.

In classical statistical physics *kinetic theory of ideal gases* is a study of systems consisting of a great number of molecules, which are considered as bodies having a small size and mass (Kikoin and Kikoin 1978). Classical statistical methods of investigation (Yavorsky and Detlaf 1975; Kikoin and Kikoin 1978; Dennery 1972; Rosser 1985; Guenault 1988; Gupta 1990; Dorlas 1999; Chandrasekhar 2000)are employed to estimate average values of quantities characterizing aggregate molecular motion such as mean velocity, mean energy, etc., which determine the macroscale characteristics of gases. The mean properties of ideal gases are calculated with the following assumptions: (1) the intramolecular forces are completely absent instead of being small; (2) the dimensions of molecules are ignored, and considered as material points; (3) the above assumptions imply the molecules are completely free, move rectilinearly and uniformly as if no forces act on them; and (4) the ceaseless chaotic movements of individual molecules obey Newton's laws of motion.

The Austrian physicist Ludwig Boltzmann suggested that knowing the probabilities for the particles to be in any of their various possible configurations would enable to work out the overall properties of the system. Going one step further, he also made a bold and insightful guess about these probabilities—that any of the many conceivable configurations for the particles would be equally probable. Boltzmann's idea works, and has enabled physicists to make mathematical models of thousands of real materials, from simple crystals to superconductors. It reflects the fact that many quantities in nature—such as the velocities of molecules in a gas—follow "normal" statistics. That is, they are closely grouped around the average value, with a "bell curve" distribution. Boltzmann's guess about equal probabilities only works for systems that have settled down to equilibrium, enjoying, for example, the same temperature throughout. The theory fails in any system where destabilizing external sources of energy are at work, such as the haphazard motion of turbulent fluids or the fluctuating energies of cosmic rays. These systems do not follow normal statistics, but another pattern instead (Buchanan 2005).

Cohen (2005) discusses Boltzmann's equation as follows. In 1872 when Boltzmann derived in his paper: *Further studies on thermal equilibrium between gas molecules* (Boltzmann 1872) what we now call the Boltzmann equation, he used, following Clausius and Maxwell, the assumption of "molecular chaos", and he does not seem to have realized the statistical, i.e., probabilistic nature of this assumption, i.e., of the assumption of the independence of the velocities of two molecules which are going to collide. He used both a dynamical and a statistical method. However, Einstein strongly disagreed with Boltzmann's statistical method, arguing that a statistical description of a system should be based on the dynamics of the system. This opened the way, especially for complex systems, for other than Boltzmann statistics. It seems that perhaps a combination of dynamics and statistics is necessary to describe systems with complicated dynamics (Cohen 2005). Sornette (2009) discusses the ubiquity of observed power law distributions in complex systems as follows. The extension of Boltzmann's distribution to out-of-equilibrium systems is the subject of intense scrutiny. In the quest to characterize complex systems, two distributions have played a leading role: the normal (or Gaussian) distribution and the power law distribution. Power laws obey the symmetry of scale invariance. Power law distributions and more generally regularly varying distributions remain robust functional forms under a large number of operations, such as linear combinations, products, minima, maxima, order statistics, powers, which may also explain their ubiquity and attractiveness. Research on the origins of power law relations, and efforts to observe and validate them in the real world, is extremely active in many fields of modern science, including physics, geophysics, biology, medical sciences, computer science, linguistics, sociology, economics and more. Power law distributions incarnate the notion that extreme events are not exceptional. Instead, extreme events should be considered as rather frequent and part of the same organization as the other events (Sornette 2009).

In the following, it is shown that the general systems theory concepts are equivalent to Boltzmann's postulates and the *Boltzmann distribution* with the inverse power law expressed as a function of the golden mean is the universal probability distribution function for the observed fractal fluctuations which corresponds closely to statistical normal distribution for moderate amplitude fluctuations and exhibit a fat long tail for hazardous extreme events in dynamical systems.

For any system large or small in thermal equilibrium at temperature T, the probability P of being in a particular state at energy E is proportional to $e^{-\frac{E}{K_B T}}$, where K_B is *Boltzmann's constant*. This is called the *Boltzmann distribution* for molecular energies and may be written as follows:

$$P \propto e^{-\frac{E}{K_B T}}. \tag{1.38}$$

The basic assumption that the space–time average of a uniform distribution of primary small scale eddies results in the formation of large eddies is analogous to Boltzmann's concept of equal probabilities for the microscopic components of the system (Buchanan 2005). The physical concepts of the general systems theory enable to derive *Boltzmann distribution* (Eq. 1.38) as shown in the following.

The r.m.s circulation speed W of the large eddy follows a logarithmic relationship with respect to the length scale ratio z equal to R/r (Eq. 1.4) as given as follows:

$$W = \frac{w_*}{k} \log z.$$

In the above equation, the variable k represents for each step of eddy growth, the fractional volume dilution of large eddy by turbulent eddy fluctuations carried on the large eddy envelope (Selvam 1990) and is given as (Eq. 1.17)

$$k = \frac{w_* r}{WR}.$$

Substituting for k in Eq. (1.4), we have

$$W = w_* \frac{WR}{w_* r} \log z = \frac{WR}{r} \log z$$

and

$$\frac{r}{R} = \log z.$$

(1.39)

The ratio r/R represents the fractional probability P of occurrence of small-scale fluctuations (r) in the large eddy (R) environment. Since the scale ratio z is equal to R/r, Eq. (1.39) may be written in terms of the probability P as follows:

$$\frac{r}{R} = \log z = \log\left(\frac{R}{r}\right) = \log\left(\frac{1}{(r/R)}\right)$$

$$P = \log\left(\frac{1}{P}\right) = -\log P.$$

(1.40)

1.7.1   General Systems Theory and Maximum Entropy Principle

The maximum entropy principle concept of classical statistical physics is applied to determine the fidelity of the inverse power law probability distribution P (Eqs. 1.16 and 1.20) for exact quantification of the observed space–time fractal fluctuations of dynamical systems ranging from the microscopic dynamics of quantum systems to macroscale real-world systems. Kaniadakis (2009) states that the correctness of an analytic expression for a given power-law tailed distribution used to describe a statistical system is strongly related to the validity of the generating mechanism. In this sense the maximum entropy principle, the cornerstone of statistical physics, is a valid and powerful tool to explore new roots in searching for generalized statistical theories (Kaniadakis 2009). The concept of entropy is fundamental in the foundation of statistical physics. It first appeared in thermodynamics through the second law of thermodynamics. In SM, we are interested in the disorder in the distribution of the system over the permissible microstates. The measure of disorder first provided by Boltzmann principle (known as Boltzmann entropy) is given by $S = K_B \ln M$, where K_B is the thermodynamic unit of measurement of entropy and is known as Boltzmann constant. $K_B = 1.33 \times 10^{-16}$ erg/°C, called thermodynamic probability or statistical weight, is the total number of microscopic complexions compatible with the macroscopic state of the system and corresponds to the "degree of disorder" or "missing information" (Chakrabarti and De 2000). For a probability distribution among a discrete set of states the generalized entropy for a system out of equilibrium is given as (Chakrabarti and De 2000; Salingaros and West 1999; Beck 2009; Sethna 2009).

$$S = -\sum_{j=1}^{\sigma} P_j \ln P_j. \tag{1.41}$$

In Eq. (1.41), P_j is the probability for the j th stage of eddy growth in this study, σ is the length step growth which is equal to the normalized deviation and the entropy S represents the "missing information" regarding the probabilities. Maximum entropy S signifies minimum preferred states associated with scale-free probabilities.

The validity of the probability distribution P (Eqs. 1.16 and 1.20) is now checked by applying the concept of maximum entropy principle (Kaniadakis 2009). Substituting for log P_j (Eq. 1.40) and for the probability P_j in terms of the golden mean τ derived earlier (Eqs. 1.16 and 1.20) the entropy S is expressed as follows:

$$S = -\sum_{j=1}^{\sigma} P_j \log P_j = \sum_{j=1}^{\sigma} P_j^2 = \sum_{j=1}^{\sigma} (\tau^{-4\sigma})^2 \tag{1.42}$$

$$S = \sum_{j=1}^{\sigma} \tau^{-8\sigma} \approx 1 \text{ for large } \sigma.$$

In Eq. (1.42), S is equal to the square of the cumulative probability density distribution and it increases with increase in σ, i.e., the progressive growth of the eddy continuum and approaches 1 for large σ. According to the second law of thermodynamics, increase in entropy signifies approach of dynamic equilibrium conditions with scale-free characteristic of fractal fluctuations and hence the probability distribution P (Eqs. 1.16 and 1.20) is the correct analytic expression quantifying the eddy growth processes visualized in the general systems theory.

Paltridge (2009) states that the principle of maximum entropy production (MEP) is the subject of considerable academic study, but is yet to become remarkable for its practical applications. The ability of a system to dissipate energy and to produce entropy "ought to be" some increasing function of the system's structural complexity. It would be nice if there were some general rule to the effect that, in any given complex system, the steady state which produces entropy at the maximum rate would at the same time be the steady state of maximum order and minimum entropy (Paltridge 2009).

Computer simulations by Damasceno, Engel, and Glotzer (2012) show that the property entropy, a tendency generally described as "disorder" can nudge particles to form organized structures. By analyzing the shapes of the particles beforehand, they can even predict what kinds of structures will form (University of Michigan 2012).

Earlier studies on the application of the concept of maximum entropy in atmospheric physics are given below. A systems theory approach based on maximum entropy principle has been applied in cloud physics to obtain useful information on droplet size distributions without regard to the details of individual droplets (Liu et al. 1995; Liu 1995; Liu and Hallett 1997, 1998; Liu and Daum 2001; Liu et al. 2002, 2002). Liu et al. (2002) conclude that a combination of the systems idea with multiscale approaches seems to be a promising avenue. Checa and Tapiador (2011)have presented a maximum entropy approach to rain drop size distribution (RDSD) modeling. Liu et al. (2011) have given a review of the concept of entropy and its relevant principles, on the organization of atmospheric systems and the principle of the second law of thermodynamics, as well as their applications to atmospheric sciences. The maximum entropy production principle (MEPP), at least as used in climate science, was first hypothesized by Paltridge (1978).

In the following it is shown that the eddy continuum energy distribution P (Eqs. 1.16 and 1.20) is the same as the *Boltzmann distribution* for molecular energies. From Eq. (1.40)

$$z = \frac{R}{r} = e^{\frac{r}{R}}$$

(1.43)

or

$$\frac{r}{R} = e^{-\frac{r}{R}}.$$

The ratio r/R represents the fractional probability P (Eqs. 1.16 and 1.20) of occurrence of small-scale fluctuations (r) in the large eddy (R) environment. Considering two large eddies of radii R_1 and R_2 (R_2 greater than R_1) and corresponding r.m.s circulation speeds W_1 and W_2 which grow from the same primary small-scale eddy of radius r and r.m.s circulation speed $w*$ we have from Eq. (1.1)

$$\frac{R_1}{R_2} = \frac{W_2^2}{W_1^2}.$$

From Eq. (1.43)

$$\frac{R_1}{R_2} = e^{-\frac{R_1}{R_2}} = e^{-\frac{W_2^2}{W_1^2}}.$$

(1.44)

The square of r.m.s circulation speed W^2 represents eddy kinetic energy. Following classical physical concepts (Kikoin and Kikoin 1978), the primary (small-scale) eddy energy may be written in terms of the eddy environment temperature T and *Boltzmann's constant* K_B as

$$W_1^2 \propto K_B T.$$

(1.44)

Representing the larger scale eddy energy as E

$$W_2^2 \propto E. \tag{1.46}$$

The length scale ratio R_1/R_2 therefore represents fractional probability P (Eqs. 1.16 and 1.20) of occurrence of large eddy energy E in the environment of the primary small-scale eddy energy $K_B T$ (Eq. 1.45). The expression for P is obtained from Eq. (1.44) as follows:

$$P \propto e^{-\frac{E}{K_B T}}. \tag{1.47}$$

The above is the same as *Boltzmann's equation* (Eq. 1.40).

The derivation of *Boltzmann's equation* from general systems theory concepts visualizes the eddy energy distribution as follows: (1) the primary small-scale eddy represents the molecules whose eddy kinetic energy is equal to $K_B T$ as in classical physics. (2) The energy pumping from the primary small-scale eddy generates growth of progressive larger eddies (Selvam 1990). The r.m.s circulation speeds W of larger eddies are smaller than that of the primary small-scale eddy (Eq. 1.1). (3) The space–time *fractal* fluctuations of molecules (atoms) in an ideal gas may be visualized to result from an eddy continuum with the eddy energy E per unit volume relative to primary molecular kinetic energy ($K_B T$) decreasing progressively with increase in eddy size.

The eddy energy probability distribution (P) of fractal space–time fluctuations also represents the *Boltzmann distribution* for each stage of hierarchical eddy growth and is given by Eqs. (1.16) and (1.20) derived earlier, namely

$$P = \tau^{-4\sigma}.$$

The general systems theory concepts are applicable to all space–time scales ranging from microscopic scale quantum systems to macroscale real-world systems such as atmospheric flows.

References

Bak PC, Tang C, Wiesenfeld K (1988) Self-organised criticality. Phys Rev A 38:364–374

Beck C (2009) Generalized information and entropy measures in physics. Contemp Phys 50:495–510 (arXiv:0902.1235v2) [cond-mat.stat-mech]

Boers R (1989) A parametrization of the depth of the entrainment zone. J Atmos Sci 28:107–111

Boltzmann L (1872) Weitere Studien Äuber das WÄarmegleichgewicht unter GasmolekÄulen. Wien Ber 66:275–370 (WA, B and I, pp 316–402)

Brown RA (1980) Longitudinal instabilities and secondary flows in the planetary boundary layer. Rev Geophys Space Phys 18:683–697

Buchanan M (2005) Entropy: the new order. New Sci 2514:34 (27 August)

Chakrabarti CG, De K (2000) Boltzmann-Gibbs entropy: axiomatic characterization and application. Int J Math Sci 23(4):243–251

Chandrasekhar BS (2000) Why are things the way they are. Cambridge University Press, Cambridge

Checa R, Tapiador FJ (2011) A maximum entropy modelling of the rain drop size distribution. Entropy 13:293–315

Cohen EGD (2005) Boltzmann and Einstein: statistics and dynamics—an unsolved problem. PRAMANA 64(5):635–643

Damasceno P, Engel M, Glotzer S (2012) Predictive self-assembly of polyhedra into complex structures. Science 337(6093):453–457

Dennery P (1972) An introduction to statistical mechanics. George Allen and Unwin, London

Devenish BJ, Bartello P, Brenguier J-L, Collins LR, Grabowski WW, IJzermans RHA, Malinowski SP, Reeks MW, Vassilicos JC, Wangi L-P, Warhaft Z (2012) Review Article: droplet growth in warm turbulent clouds. Quart J R Meteor Soc 138:1401–1429, (July 2012 B)

Donelly RJ (1988) Superfluid turbulence. Sci Am 259(5):100–109

Donelly RJ (1990) Quantized vortices in helium II. Cambridge University Press, USA

Dorlas TC (1999) Statistical mechanics. Institute of Physics Publishing, Bristol

Eady ET (1950) The cause of the general circulation of the atmosphere. Cent Proc Roy Met Soc 156–172

Grabowskii WW, Wang L-P (2013) Growth of cloud droplets in a turbulent environment. Annu Rev Fluid Mech 45:293–324

Gryning S-E, Batchvarova E (2006) Parametrization of the depth of the entrainment zone above the daytime mixed layer. Quart J R Meteor Soc 120(515):47–58

Guenault M (1988) Statistical Physics. Routledge, London

Gupta MC (1990) Statistical thermodynamics. Wiley Eastern, New Delhi

Haken H (1977) Synergetics, an introduction: nonequilibrium phase transitions and self-organization in physics, chemistry, and biology. Springer, New York

Hultmark M, Vallikivi M, Bailey S, Smits A (2012) Turbulent pipe flow at extreme Reynolds numbers. Phys Rev Lett 108(9):094501

Holton JR (2004) An introduction to dynamic meteorology. Academic, USA

Kaniadakis G (2009) Maximum entropy principle and power-law tailed distributions. Eur Phys J B 70:3–13

Kikoin AK, Kikoin IK (1978) Molecular physics. Mir, Moscow

Klir GJ (2001) Facets of systems science, 2nd edn (IFSR International Series on Systems Science and Engineering vol 15). Kluwer Academic/Plenum, New York

Lawrence Berkeley National Laboratory (2013) How computers push on the molecules they simulate. retrieved 28 December 2014 from http://phys.org/news/2013-01-molecules-simulate.html

Lebowitz JL (1999) Statistical mechanics: a selective review of two central issues. Rev Modern Phys 71:346–S357 (Reprinted in "More things in heaven and earth: a celebration of physics at the millenium," (ed. B. Bederson) Springer, (1999), pp 581–600. http://arxiv.org/abs/math-ph/0010018v1)

Liu V (1956) Turbulent dispersion of dynamic particles. J Meteor 13:399–405

Liu Y (1992) Skewness and kurtosis of measured raindrop size distributions. Atmos Environ 26A:2713–2716

Liu Y (1995) On the generalized theory of atmospheric particle systems. Adv Atmos Sci 12:419–438

Liu Y, Daum PH (2001) Statistical physics, information theory and cloud droplet size distributions. Eleventh ARM Science Team Meeting Proceedings, Atlanta, Georgia, March 19–23

Liu Y, Hallett J (1997) The "1/3" power-law between effective radius and liquid water content. Quart J Roy Meteor Soc 123:1789–1795

Liu Y, Hallett J (1998) On size distributions of droplets growing by condensation: a new conceptual model. J Atmos Sci 55:527–536

Liu Y, Laiguang Y, Weinong Y, Feng L (1995) On the size distribution of cloud droplets. Atmos Res 35:201–216

Liu Y, Daum PH, Hallett J (2002) A generalized systems theory for the effect of varying fluctuations on cloud droplet size distributions. J Atmos Sci 59:2279–2290

Liu Y, Daum PH, Chai SK, Liu F (2002) Cloud parameterizations, cloud physics, and their connections: an overview. Recent Res Devel Geophys 4:119–142

Liu Y, Liu C, Wang D (2011) Understanding atmospheric behaviour in terms of entropy: a review of applications of the second law of thermodynamics to meteorology. Entropy 13:211–240

Lovejoy S, Schertzer D (2010) Towards a new synthesis for atmospheric dynamics: space–time cascades. Atmos Res 96:1–52

Marusic I, McKeon BJ, Monkewitz PA, Nagib HM, Smits AJ, Sreenivasan KR (2010) Wall-bounded turbulent flows at high Reynolds numbers: recent advances and key issues. Phys Fluids 22:065103

McGraw R, Liu Y (2003) Kinetic potential and barrier crossing: a model for warm cloud drizzle formation. Phys Rev Lett 90(1):018501

Newman MEJ (2011) Complex systems: a survey. arXiv:1112.1440v1 [cond-mat.stat-mech]

Paltridge GW (1978) Climate and thermodynamic systems of maximum dissipation. Nature 279:630–631

Paltridge GW (2009) A story and a recommendation about the principle of maximum entropy production. Entropy 11:945–948

Parmeggiani A (2012) Viewpoint. Physics 5:118 (Gorissen M, Lazarescu A, Mallick K, Vanderzande C (2012) A viewpoint on: exact current statistics of the asymmetric simple exclusion process with open boundaries. Phys Rev Lett 109:170601)

Peters O, Hertlein C, Christensen K (2002) A complexity view of rainfall. Phys Rev Lett 88:018701

Peters O, Deluca A, Corral A, Neelin JD, Holloway CE (2010) Universality of rain event size distributions. J Stat Mech November: P11030

Prandtl L (1932) Zur turbulenten Strömung in Rohren und längs Platten. Ergebnisse der Aerodynamischen Versuchsanstalt zu Göttingen 4:18–29

Pruppacher HR, Klett JD (1997) Microphysics of clouds and precipitation. Kluwer Academic, The Netherlands

Rosser WG (1985) An introduction to statistical physics. Ellis Horwood, Chichester

Salingaros NA, West BJ (1999) A universal rule for the distribution of sizes. Environ Plan B: Plan Des 26:909–923 (Pion Publications)

Selvam AM (1990) Deterministic chaos, fractals and quantumlike mechanics in atmospheric flows. Can J Phys 68:831–841. retrieved 28 December 2014 from http://xxx.lanl.gov/html/physics/0010046

Selvam AM (1993) Universal quantification for deterministic chaos in dynamical systems. Appl Math Model 17:642–649. retrieved 28 December 2014 from http://xxx.lanl.gov/html/physics/0008010

Selvam AM (2005) A general systems theory for chaos, quantum mechanics and gravity for dynamical systems of all space–time scales. Electromagn Phenom 5(15):160–176. retrieved 28 December 2014 from http://arxiv.org/pdf/physics/0503028; http://www.emph.com.ua/15/selvam.htm

Selvam AM (2007) Chaotic climate dynamics. Luniver, U.K.

Selvam AM (2009) Fractal fluctuations and statistical normal distribution. Fractals 17(3):333–349. retrieved 28 December 2014 from http://arxiv.org/pdf/0805.3426

Selvam AM (2011) Signatures of universal characteristics of fractal fluctuations in global mean monthly temperature anomalies. J Syst Sci Complex 24(1):14–38. retrieved 28 December 2014 from http://arxiv.org/abs/0808.2388.

Selvam AM (2012a) Universal spectrum for atmospheric suspended particulates: comparison with observations. Chaos Complex Lett 6(3):1–43. retrieved 28 December 2014 from http://arxiv.org/abs/1005.1336

Selvam AM (2012b) Universal spectrum for atmospheric aerosol size distribution: comparison with pcasp-b observations of vocals 2008. Nonlinear Dyn Syst Theory 12(4):397–434. retrieved 28 December 2014 from http://arxiv.org/abs/1105.0172

Selvam AM (2013) Scale-free universal spectrum for atmospheric aerosol size distribution for Davos, Mauna Loa and Izana. Int J Bifurcation Chaos 23(1350028):1–13

Selvam AM (2014) A general systems theory for rain formation in warm clouds. Chaos Complex Lett 8(1):1–42. retrieved 28 December 2014 from http://arxiv.org/pdf/arxiv1211.0959

Selvam AM, Fadnavis S (1998) Signatures of a universal spectrum for atmospheric inter-annual variability in some disparate climatic regimes. Meteor Atmos Phys 66:87–112. retrieved 28 December 2014 from http://xxx.lanl.gov/abs/chao-dyn/9805028

Sethna JP (2009) Statistical mechanics: entropy, order parameters, and complexity. Clarendon Press, Oxford. retrieved 28 December 2014 from http://www.freebookcentre.net/physics-books-download/Statistical-Mechanics-Entropy,-Order-Parameters,-and-Complexity-[PDF-371].html

Sivak DA, Chodera JD, Crooks GE (2013) Using nonequilibrium fluctuation theorems to understand and correct errors in equilibrium and nonequilibrium discrete Langevin dynamics simulations. Phys. Rev. doi:10.1103/PhysRevX.3.011007

Skyttner L (2005) General systems theory: problems, perspectives, practice, 2nd edn. World Scientific, Singapore

Sornette D (2009) Probability distributions in complex systems. arXiv:0707.2194v1 [physics.data-an] (2009). This is the Core article for the Encyclopedia of Complexity and System Science, Springer Science

Townsend AA (1956) The structure of turbulent shear flow, 2nd edn. Cambridge University Press, London, pp 115–130

Tuck AF (2010) From molecules to meteorology via turbulent scale invariance. Quart J R Meteor Soc 136:1125–1144

University of Michigan (2012) Entropy can lead to order, paving the route to nanostructures. ScienceDaily. retrieved 28 December 2014 from http://www.sciencedaily.com/releases/2012/07/120726142200.htm.

Von Bertalanffy L (1972) The history and status of general systems theory. Acade Manag J 15:407–426 (General Systems Theory)

Von Karman T (1956) Mechanische Ähnlichkeit und Turbulenz, Nachr. Ges. Wiss. Gött., Math.-phys. Kl. 58–76

Yano J-I, Liu C, Moncrieff MW (2012) Self-organised criticality and homeostasis in atmospheric convective organization. J Atmos Sci 69:3449–3462

Yavorsky B, Detlaf A (1975) Handbook of physics. Mir, Moscow

Chapter 2
Cumulus Cloud Model

Abstract A cumulus cloud model that can explain the observed characteristics of warm rain formation in monsoon clouds is presented. The model is based on classical statistical physical concepts and satisfies the principle of maximum entropy production. Atmospheric flows exhibit self-similar fractal fluctuations that are ubiquitous to all dynamical systems in nature and are characterized by inverse power-law form for power (eddy energy) spectrum signifying long-range space–time correlations. A general systems theory model for atmospheric flows is based on the concept that the large eddy energy is the integrated mean of enclosed turbulent (small-scale) eddies. This model gives scale-free universal governing equations for cloud-growth processes. The model-predicted cloud parameters are in agreement with reported observations, in particular, the cloud drop-size distribution. Rain formation can occur in warm clouds within a 30-min lifetime under favourable conditions of moisture supply in the environment.

Keywords General systems theory · Nonlinear dynamics and chaos · Fractals · Long-range space–time correlations · Inverse power-law eddy energy spectrum · Maximum entropy production principle

2.1 Introduction

The knowledge of the cloud dynamical, microphysical and electrical parameters and their interactions are essential for the understanding of the formation of rain in warm clouds and their modification. Extensive aircraft observations of cloud dynamical, microphysical and electrical parameters have been made in more than 2000 isolated warm cumulus clouds formed during the summer monsoon seasons (June–September) in Pune (18°32′N, 73°51′E, 559 m a.s.l), India (Selvam et al. 1980, 1982a, 1982b, 1982c, 1982d, 1983, 1984a, 1984b, 1984c; Murty et al. 1985; Selvam et al. 1991a, 1991b). The observations were made during aircraft traverses at about 300 m above the cloud base. These observations have provided new evidence relating to the dynamics of monsoon clouds. A brief summary of the important results is given as follows: (i) Horizontal structure of the air flow inside the cloud has consistent variations with successive positive and negative values of vertical velocity representative of ascending and descending air currents inside the

© The Author(s) 2015

A. M. Selvam, *Rain Formation in Warm Clouds*, SpringerBriefs in Meteorology,
DOI 10.1007/978-3-319-13269-3_2

cloud. (ii) Regions of ascending currents are associated with higher liquid water content (LWC), and negative cloud drop charges and the regions of descending current are associated with lower LWC and positive cloud drop charges. (iii) Width of the ascending and descending currents is about 100 m. The ascending and descending currents are hypothesized to be due to cloud-top-gravity oscillations (Selvam et al. 1982a, 1982b; 1983). The cloud-top-gravity oscillations are generated by the intensification of turbulent eddies due to the buoyant production of energy by the microscale fractional condensation (MFC) in turbulent eddies. (iv) Measured LWC (q) at the cloud-base levels is smaller than the adiabatic value (q_a) with $q/q_a = 0.6$. The LWC increases with height from the base of the cloud and decreases towards the cloud-top regions. (v) Cloud electrical activity is found to increase with the cloud LWC. (vi) Cloud-drop spectra are unimodal near the cloud base and multimodal at higher levels. The variations in mean volume diameter (MVD) are similar to those in the LWC. (vii) In-cloud temperatures are colder than the environment. (viii) The lapse rates of the temperatures inside the cloud are less than the immediate environment. Environmental lapse rates are equal to the saturated adiabatic value. (ix) Increments in the LWC are associated with increments in the temperature inside the cloud. The increments in temperature are associated with the increments in temperature of the immediate environment at the same level or the level immediately above. (x) Variances of in-cloud temperature and humidity are higher in the regions where the values of LWC are higher (Selvam et al. 1982a, 1982b, 1982c, 1982d). The variances of temperature and humidity are larger in the clear-air environment than in the cloud air (Selvam et al. 1982a, 1982b, 1982c, 1982d).

The dynamical and physical characteristics of monsoon clouds described above cannot be explained by simple entraining cloud models. A simple cumulus cloud model, which can explain the observed cloud characteristics, has been developed (Selvam et al. 1983). The relevant physical concept and theory relating to the dynamics of atmospheric planetary boundary layer (PBL), formation of warm cumulus clouds and their modification through hygroscopic particle seeding are presented in the following sections.

The mechanism of large eddy growth, discussed in Sect. 2.4, in the atmospheric ABL can be applied to the formulation of the governing equations for cumulus cloud growth. Based on the above theory, equations are derived for the in-cloud vertical profiles of (i) ratio of actual cloud LWC (q) to the adiabatic LWC (q_a), (ii) vertical velocity, (iii) temperature excess, (iv) temperature lapse rate, (v) total LWC (q_t), (vi) cloud growth time, (vii) cloud drop-size spectrum, and (viii) raindrop size spectrum. The equations are derived starting from the MFC process at cloud-base levels. This provides the basic energy input for the total cloud growth.

2.1.1   Vertical Profile of q/q_a

The observations of cloud LWC, q, indicate that the ratio q/q_a is less than 1 due to dilution by vertical mixing. The fractional volume dilution rate f in the cloud updraft can be computed (Selvam et al. 1983; Selvam et al. 1984a; Selvam et al. 1984b, Selvam 1990, 2007) from Eq. (8) (see Sect. 1.5.3) given by

$$f = \sqrt{\frac{2}{\pi z}} \ln z.$$

In the above equation, f represents the fraction of the air mass of the surface origin which reaches the height z after dilution by vertical mixing caused by the turbulent eddy fluctuations.

Considering that the cloud-base level is 1000 m, the value of $R = 1000$ m and the value of turbulence length scale r below cloud base is equal to 100 m so that the normalized length scale $z = R/r = 1000$ m/100 m $= 10$, and the corresponding fractional volume dilution $f = 0.6$.

The value of q/q_a at the cloud-base level is also found to be about 0.6 by several observers (Warner 1970).

The fractional volume dilution f will also represent the ratio q/q_a inside the cloud. The observed (Warner 1970) q/q_a profile inside the cloud is seen (closely) to follow the profile obtained by the model for dominant eddy radius $r = 1$ m (Fig. 1.4). It is, therefore, inferred that, inside the cloud, the dominant turbulent eddy radius is 1 m, while below the cloud base, the dominant turbulent eddy radius is 100 m.

2.1.2 In-Cloud Vertical Velocity Profile

The logarithmic wind-profile relationship (Eq. 1.4) derived for the PBL in Sect. 1.5.2 holds good for conditions inside a cloud because the same basic physical process, namely MFC, operates in both the cases. The value of vertical velocity inside the cloud will, however, be much higher than in cloud-free air.

From Eq. (1.6), the in-cloud vertical velocity profile can be expressed as

$$W = w_* fz,$$

where W is the vertical velocity at height z, w_* is the production of vertical velocity per second by the MFC at the reference level, i.e. cloud-base level, and f is the fractional upward mass flux of air at level z originating from the cloud-base level.

The f profile is shown in Fig. 1.4. The vertical velocity profile will follow the fz profile assuming w_* is the constant at the cloud-base level during the cloud-growth period.

2.1.3 In-Cloud Excess Temperature Perturbation Profile

The relationship between temperature perturbation θ and the corresponding vertical velocity perturbation is given as follows:

$$W = \frac{g}{\theta_0} \theta,$$

where g is the acceleration due to gravity and θ_0 is the reference-level potential temperature at the cloud-base level.

By substituting for W and taking θ_* as the production of temperature perturbation at the cloud-base level by MFC, we arrive at the following expression since there is a linear relationship between the vertical velocity perturbation W and temperature perturbation θ (from Eqs. 1.4 and 1.6):

$$\theta = \frac{\theta_*}{k} \ln z = \theta_* fz. \tag{2.1}$$

Thus, the in-cloud vertical velocity and temperature perturbation follow the fz distribution (Fig. 1.5).

2.1.4 In-Cloud Temperature Lapse Rate Profile

The saturated adiabatic lapse rate Γ_{sat} is expressed as

$$\Gamma_{sat} = \Gamma - \frac{L}{C_p} \frac{d\chi}{dz},$$

where Γ is the dry adiabatic lapse rate, C_p is the specific heat of air at constant pressure, and $d\chi/dz$ is the liquid water condensed during parcel ascent along a saturated adiabat Γ_{sat} in a height interval dz.

In the case of cloud growth with vertical mixing, the in-cloud lapse rate Γ_s can be written as

$$\Gamma_s = \Gamma - \frac{L}{C_p} \frac{dq}{dz},$$

where dq, which is less than $d\chi$, is the liquid water condensed during a parcel ascent dz and q is less than the adiabatic LWC q_a. From Eq. (2.1),

$$\Gamma_s = \Gamma - \frac{d\theta}{dz} = \Gamma - \frac{\theta}{r} = \Gamma - \frac{\theta_* fz}{r}, \tag{2.2}$$

where $d\theta$ is the temperature perturbation θ during parcel ascent dz. By concept, dz is the dominant turbulent eddy radius r (Fig. 1.2).

2.1.5 Total Cloud LWC Profile

The total cloud LWC q_t at any level is directly proportional to θ as given by the following expression:

$$q_t = \frac{C_p}{L}\theta = \frac{C_p}{L}\theta_* fz = q_* fz, \tag{2.3}$$

where q_* is the production of LWC at the cloud-base level and is equal to $C_p\theta_*/L$. The total cloud LWC q_t profile follows the fz distribution (Fig. 1.5).

2.1.5.1 Cloud-Growth Time

The large eddy-growth time (Eq. 1.36) can be used to compute cloud-growth time T_c:

$$T_c = \frac{r_*}{w_*}\sqrt{\frac{\pi}{2}}li(\sqrt{z})_{z_1}^{z_2}, \tag{2.4}$$

where li is the Soldner's integral or the logarithm integral. The cloud growth time T_c using Eq. (2.4) is shown in Fig. 2.1.

2.2 Cloud Model Predictions and Comparison with Observations

Numerical computations of cloud parameters were performed for two different cloud-base cloud condensation nuclei (CCN) mean volume radii, namely 2.2 and 2.5 μm, and computed values are compared with the observations. The results are discussed below.

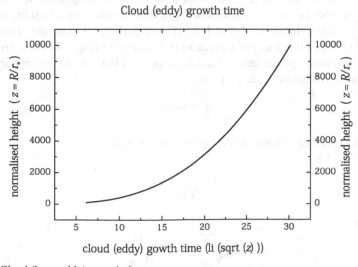

Fig. 2.1 Cloud (large eddy) growth time

2.2.1   Vertical Velocity Profile in the Atmospheric
Boundary Layer

The MFC-generated values of vertical velocity have been calculated for different heights above the surface for clear air conditions and above the cloud base for in-cloud conditions for a representative tropical environment with favourable moisture supply. A representative cloud-base height is considered to be 1000 m above sea level (a.s.l), and the corresponding meteorological parameters are surface pressure 1000 mb, surface temperature 30 °C, relative humidity at the surface 80 % and turbulent length scale 1 cm. The values of the latent heat of vaporisation L_V and the specific heat of air at constant pressure C_p are 600 and 0.24 cal gm^{-1}, respectively. The ratio values of m_w/m_0, where m_0 is the mass of the hygroscopic nuclei per unit volume of air and m_w is the mass of water condensed on m_0, at various relative humidities as given by Winkler and Junge (1971, 1972) have been adopted and the value of m_w/m_0 is equal to about 3 for relative humidity of 80 %. For a representative value of m_0 equal to 100 μg m^{-3}, the temperature perturbation θ' is equal to 0.00065 °C, and the corresponding vertical velocity perturbation (turbulent) w_* is computed and is equal to 21.1×10^{-4} cm s^{-1} from the following relationship between the corresponding virtual potential temperature θ_v and the acceleration due to gravity g, which is equal to 980.6 cm s^{-2}:

$$w_* = \frac{g}{\theta_v}\theta'.$$

Heat generated by the condensation of water equal to 300 μg on 100 μg of hygroscopic nuclei per metre cube, say in 1 s, generates vertical velocity perturbation w_* (cm s^{-2}) equal to 21.1×10^{-4} cm s^{-2} at surface levels. Since the time duration for water vapour condensation by deliquescence is not known, in the following it is shown that a value of w_* equal to 30×10^{-7} cm s^{-2}, i.e. about three orders of magnitude less than that shown in the above example is sufficient to generate clouds as observed in practice.

From the logarithmic wind-profile relationship (Eq. 1.4) and the steady state fractional upward mass flux f of surface air at any height z (Eq. 1.8), the corresponding vertical velocity perturbation W can be expressed in terms of the primary vertical velocity perturbation w_* as (Eq. 1.6):

$$W = w_* fz,$$

W may be expressed in terms of the scale ratio z as follows:
From Eq. (1.8),

$$f = \sqrt{\frac{2}{\pi z}} \ln z.$$

Therefore,

$$W = w_* z \sqrt{\frac{2}{\pi z}} \ln z = w_* \sqrt{\frac{2z}{\pi}} \ln z.$$

Table 2.1 Vertical profile of eddy vertical velocity perturbation W

Height above surface R	Length scale ratio $z = R/r_*$	Vertical velocity $W = w_* fz$ cm s^{-1}
1 cm	1 ($r_* = 1$ cm)	$30 \times 10^{-7} (= w_*)$
100 cm	100	1.10×10^{-4}
100 m	100×100	2.20×10^{-3}
1 km	1000×100	$8.71 \times 10^{-3} \approx 0.01$
10 km	10000×100	3.31×10^{-2}

The values of large eddy vertical velocity perturbation W produced by the process of MFC at normalized height z computed from Eq. 1.6 are given in Table 2.1. The turbulence length scale r_* is equal to 1 cm, and the related vertical velocity perturbation w_* is equal to 30×10^{-7} cm/s for the height interval 1 cm to 1000 m (cloud-base level) for the computations shown in Table 2.1. Progressive growth of successively larger eddies generates a continuous spectrum of semipermanent eddies anchored to the surface and with increasing circulation speed W.

The above values of vertical velocity, although small in magnitude, are present for long enough time period in the lower levels and contribute to the formation and development of clouds as explained in the next section.

2.2.2   Large Eddy-Growth Time

The time T required for the large eddy of radius R to grow from the primary turbulence scale radius r_* is computed from Eq. (1.36) as follows:

$$T = \frac{r_*}{w_*} \sqrt{\frac{\pi}{2}} \int_{x1}^{x2} \mathrm{li}(\sqrt{z}).$$

$$x_1 = \sqrt{z_1} \text{ and } x_2 = \sqrt{z_2}.$$

In the above equation, z_1 and z_2 refer, respectively, to the lower and upper limits of integration and li is the Soldner's integral or the logarithm integral. The large eddy-growth time T can be computed from Eq. (1.36) as follows.

As explained earlier, a continuous spectrum of eddies with progressively increasing speed (Table 2.1) anchored to the surface grows by MFC originating in turbulent fluctuations at the planetary surface. The eddy of radius 1000 m has a circulation speed equal to 0.01 cm/s (Table 2.1). The time T seconds taken for the evolution of the 1000-m (10^5 cm) eddy from 1 cm height at the surface can be computed from the above equation by substituting for $z_1 = 1$ cm and $z_2 = 10^5$ cm such that $x_1 = 1$ and $x_2 \approx 317$.

$$T = \frac{1}{0.01} \sqrt{\frac{\pi}{2}} \int_{1}^{317} \mathrm{li}(z).$$

The value of $\int_1^{317} \mathrm{li}(z)$ is equal to 71.3.

Hence, $T \approx 8938$, $s \approx 2$ h 30 min.

Thus, starting from the surface level cloud growth begins after 2 h 30 min. This is consistent with the observations that under favourable synoptic conditions solar surface heating during the afternoon hours gives rise to cloud formation.

The dominant turbulent eddy radius at 1000 m in the sub-cloud layer is 100 m starting from the 1-cm-radius dominant turbulent eddy at surface and the formation of successively larger dominant eddies at decadic length scale intervals as explained in Sect. 1.5.1. Also, it has been shown in Sect. 2.1.1 that the radius of the dominant turbulent eddy (r_*) inside the cloud is 1 m. These features suggest that the scale ratio is 100 times larger inside the cloud than below the cloud. The 1000-m (1 km) eddy at cloud-base level forms the internal circulation for the next stage of eddy growth, namely 10 km eddy radius with circulation speed equal to 0.03 cm/s. Cloud growth begins at 1 km above the surface and inside this 10-km eddy, with dominant turbulent eddy radius 1 m as shown above. The circulation speed of this 1-m-radius eddy inside cloud is equal to 3 m/s as shown in the following. Since the eddy continuum ranging from 1 cm to 10 km radius grows from the surface starting from the same primary eddy of radius r_* and the perturbation speed w_* (cm/s), the circulation speeds of any two eddies of radii R_1, R_2 with corresponding circulation speeds W_1 and W_2 are related to each other as follows from Eq. (1.1):

$$W_1^2 = \sqrt{\frac{2}{\pi} \frac{r_*}{R_1}} \, w_*^2,$$

$$W_2^2 = \sqrt{\frac{2}{\pi} \frac{r_*}{R_2}} \, w_*^2,$$

$$\frac{W_2^2}{W_1^2} = \frac{R_1}{R_2},$$

$$\frac{W_2}{W_1} = \sqrt{\frac{R_1}{R_2}}.$$

As mentioned earlier, cloud growth with dominant turbulent eddy radius 1 m begins at 1 km above surface and forms the internal circulation to the 10-km eddy. The circulation speed of the in-cloud dominant turbulent eddy is computed as equal to 3 m/s from the above equation where the subscripts 1 and 2 refer, respectively, to the outer 10-km eddy and the internal 1-m eddy.

The value of vertical velocity perturbation W at cloud base is then equal to 100 times the vertical velocity perturbation just below the cloud base. Vertical velocity perturbation just below the cloud base is equal to 0.03 cm/s from Table 2.1. Therefore, the vertical velocity perturbation at cloud base is equal to 0.03×100 cm/s, i.e. 3 cm/s and is consistent with airborne observations over the Indian region during the monsoon season (Selvam et al. 1976; Pandithurai et al. 2011).

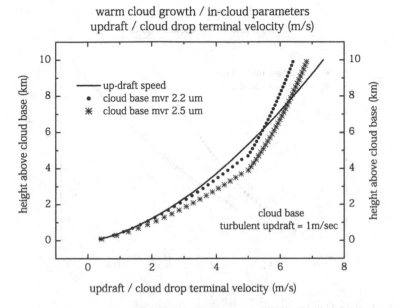

Fig. 2.2 In-cloud updraft speed and cloud particle terminal velocities for the two input cloud-base CCN size spectra with mean volume radius (mvr) equal to (i) 2.2 μm and (ii) 2.5 μm. CCN cloud condensation nuclei

Cloud-base vertical velocity equal to 1 cm/s has been used for the model computations in the following. The in-cloud updraft speed and cloud particle terminal velocities are given in Fig. 2.2 for the two input cloud-base CCN size spectra with mean volume radius (mvr) equal to (i) 2.2 μm and (ii) 2.5 μm. The in-cloud updraft speed W is the same for both CCN spectra since $W = w_* fz$ (Eq. 1.6) and depends only on the persistent cloud-base primary perturbation speed w_* originating from MFC by deliquescence on hygroscopic nuclei at surface levels in humid environment (see Sect. 1.3). Cloud LWC increases with height (Fig. 2.3) associated with the increase in cloud particle mean volume radius (Fig. 2.4) and terminal velocities (Fig. 2.2). The cloud particles originating from the larger size CCN (mvr=2.5 μm) are associated with larger cloud LWCs, larger mean volume radii, and, therefore, larger terminal fall speeds at all levels.

The turbulent vertical velocity perturbation w_* at cloud-base level (1 km) is equal to 0.01 m/s or 1 cm/s. The corresponding cloud-base temperature perturbation θ_* is then computed from the equation:

$$w_* = \frac{g}{\theta_v} \theta_*$$

$$\theta_* = w_* \frac{\theta_v}{g}.$$

Substituting $w_*=1$ cm/s, $\theta_v=273+30=303$ K and $g=980.6$ cm/s^2, the temperature perturbation is equal to 0.309 °C and for the 1-m eddy radius the average in-cloud

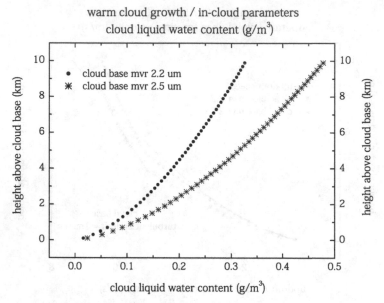

Fig. 2.3 Cloud liquid water content

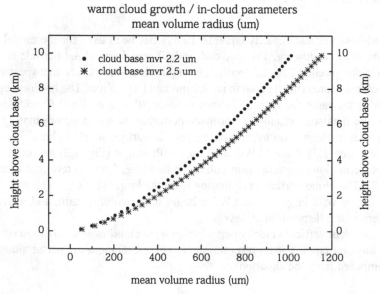

Fig. 2.4 Cloud particle mean volume radius (μm)

temperature perturbation per centimetre is equal to $0.309/100 = 0.00309\,°C$. The temperature perturbation (warming) $\theta = \theta_s fz$ (Eq. 1.6) increases with height with corresponding decrease in the in-cloud temperature lapse.

2.2.3   In-Cloud Temperature Lapse Rate

The in-cloud lapse rate Γ_s is computed using the following expression (Eq. 2.2):

$$\Gamma_s = \Gamma - \frac{\theta_* fz}{r_*}.$$

The primary eddy radius length r_* at cloud base is equal to 1 m as shown earlier (Sect. 2.1). The model computations for in-cloud vertical profile of vertical velocity W, temperature perturbation θ and lapse rate Γ_c at 1 km height intervals above the cloud base are given in Table 2.2.

The predicted temperature lapse rate decreases with height and becomes less saturated than adiabatic lapse rate near the cloud top, the in-cloud temperatures being warmer than the environment. These results are in agreement with the observations.

2.2.4   Cloud-Growth Time

The large eddy-growth time (Eq. 1.36) can be used to compute cloud-growth time T_c (Eq. 2.4):

$$T_c = \frac{r_*}{w_*}\sqrt{\frac{\pi}{2}}\mathrm{li}(\sqrt{z})_{z_1}^{z_2},$$

where li is the Soldner's integral or the logarithm integral. The vertical profile of cloud-growth time T_c is a function of the cloud-base primary turbulent eddy fluctuations of radius r_* and perturbation speed w_* alone. The cloud-growth time T_c using Eq. (2.4) is shown in Fig. 2.5 for the two different cloud-base CCN spectra, with mean volume radii equal to 2.2 and 2.5 μm, respectively. The cloud-growth time remains the same since the primary trigger for cloud growth is the persistent turbulent energy generation by condensation at the cloud base in primary turbulent eddy fluctuations of radius r_* and perturbation speed w_*.

Let us consider r_* is equal to 100 cm and w_* is equal to 1 cm s^{-1} the time taken for the cloud to grow (see Sect. 2.1.6) to a height of, e.g. 1600 m above cloud base can be computed as shown below. The normalized height z is equal to 1600 since dominant turbulent eddy radius is equal to 1 m:

$$T_c = \frac{100}{100 \times 0.01}\sqrt{\frac{\pi}{2}}\mathrm{li}\sqrt{1600}$$
$$= 100 \times 1.2536 \times 15.84 \ s$$
$$\approx 30 \ \mathrm{min}.$$

The above value is consistent with cloud-growth time observed in practice.

Table 2.2 In-cloud vertical velocity and temperature lapse rates

Sl. No.	Height above cloud base R(m)	Scale ratio $z = R/r_*$, ($r_* = 1$ m)	f	fz	In-cloud vertical velocity (ms^{-1}) $W = w_* fz$ ms^{-1}, $w_* = .01$ ms^{-1}	In-cloud temperature perturbation C $\theta = (\theta_*, fz)$ C $\theta_* = 0.00309$C	In-cloud lapse rate Γ_c°C/km $\Gamma_c' = \Gamma - \theta$ $\Gamma = -10$ C/km
1	1000	1000	0.17426	174.26	1.74	0.538	−9.46
2	2000	2000	0.13558	271.16	2.71	0.838	−9.16
3	3000	3000	0.11661	349.82	3.50	1.081	−8.92
4	4000	4000	0.10461	418.46	4.18	1.293	−8.70
5	5000	5000	0.09609	480.43	4.80	1.484	−8.52
6	6000	6000	0.08959	537.56	5.38	1.661	−8.34
7	7000	7000	0.08442	590.91	5.91	1.826	−8.17
8	8000	8000	0.08016	641.24	6.41	1.981	−8.01
9	9000	9000	0.07656	689.05	6.89	2.129	−7.87
10	10,000	10,000	0.07347	734.73	7.34	2.270	−7.72

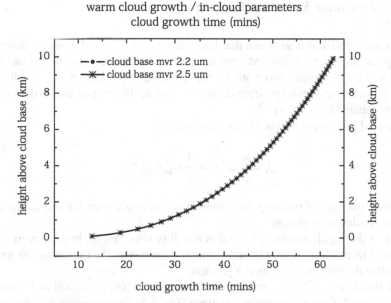

Fig. 2.5 Cloud-growth time

2.2.5 *Cloud Drop-Size Spectrum*

The evolution of cloud drop-size spectrum is critically dependent on the water vapour available for condensation and the nuclei number concentration in the sub-cloud layer. Cloud drops form as water vapour condenses in the air parcel ascending from cloud-base levels. Vertical mixing during ascent reduces the volume of cloud-base air reaching higher levels to a fraction f of its initial volume. Thus, the nuclei available for condensation, i.e. the cloud drops number concentration also decreases with height according to the f distribution. The total cloud-water content was earlier shown (Eq. 1.6) to increase with height according to the fz distribution. Thus, bigger size cloud drops are formed on the lesser number of condensation nuclei available at higher levels in the cloud. Due to vertical mixing, unsaturated conditions exist inside the cloud. Water vapour condenses on fresh nuclei available at each level, since, in the unsaturated in-cloud conditions, MFC occurs preferentially on small condensation nuclei (Pruppacher and Klett 1997).

Earlier in Sect. 1.6, it was shown that the atmospheric eddy continuum fluctuations hold in suspension atmospheric particulates, namely aerosols, cloud droplets and raindrops. The cloud drop-size distribution spectrum also follows the universal spectrum derived earlier for atmospheric aerosol size distribution.

2.2.6 In-Cloud Raindrop Spectra

In the cloud model, it is assumed that bulk conversion of cloud water to rainwater takes place mainly by collection, and the process is efficient due to the rapid increase in the cloud-water flux with height. The in-cloud raindrop size distribution also follows the universal spectrum derived earlier for the suspended particulates in the atmosphere (Sect. 1.6.4).

The total rainwater content Q_r (c.c) is given as

$$Q_r = \frac{4}{3}\pi r_a^3 N = \frac{4}{3}\pi r_{as}^3 N_* f z^2.$$

(2.5)

The above concept of raindrop formation is not dependent on the individual drop collision coalescence process.

Due to the rapid increase of cloud water flux with height, bulk conversion to rain water takes place in time intervals much shorter than the time required for the conventional collision–coalescence process.

The cloud-base CCN size spectrum and the in-cloud particle (cloud and raindrop) size spectrum follow the universal spectrum (Fig. 1.8) for suspended particulates in turbulent fluid flows. The cloud-base CCN spectra and the in-cloud particulates (cloud and raindrops) size spectra at two levels, 100 m and 2 km, plotted in conventional manner as dN/Nd (log R) versus R on log–log scale are shown in Fig. 2.6. The in-cloud particulate size spectrum shifts rapidly towards larger sizes associated with

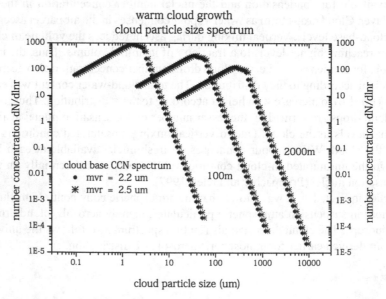

Fig. 2.6 Cloud-base CCN spectra and the in-cloud particulates (cloud and raindrops) size spectra at two levels—100 m and 2 km. CCN cloud condensation nuclei

rain formation. According to the general systems theory for suspended particulate size spectrum in turbulent fluid flows (Sect. 1.6), larger suspended particulates are associated with the turbulent regions (smaller scale length with larger fluctuation speed) of the vertical velocity spectrum. Spontaneous formation of larger cloud/raindrops may occur by collision and coalescence of smaller drops in these regions of enhanced turbulence.

2.2.7 Rainfall Commencement

Rainfall sets in at the height at which the terminal velocity w_T of the raindrop becomes equal to the mean cloud updraft W. Let the mean volume radius R_m be representative of the precipitation drop at level z above cloud base. In the intermediate range (40 µm $< R_m <$ 0.6 mm), an approximated formula for the fall speed is given by Yau and Rogers (1989) as $w_T = K_3 R_m$, where $K_3 = 8 \times 10^3$ s^{-1}.

For droplets with radii $<$30 µm, the terminal velocity is given as $w_T = K_1 R_m^2$, where $K_1 \sim 1.19 \times 10^6$ cm^{-1} s^{-1}.

For large drops, $w_T = K_2 \sqrt{R_m}$. K_2 can be approximated as $K_2 = 2010$ cm$^{1/2}$ s^{-1} for a raindrop size between 0.6 and 2 mm (Yau and Rogers 1989).

2.2.8 Rainfall Rate

The in-cloud rainfall rate (R_t) can be computed as shown below.

Rain water production rate over unit area (rainfall rate mm/s/unit area) Q_{rt} in the cloud at any level z above the cloud base (Eq. 2.5) is given by

$$R_{tz} = (W_T - W)\frac{4}{3}\pi R_m^3 N.$$

The in-cloud rainfall rate R_{tz} is derived from the normalized flux value for height z, i.e. eddy wavelength z. Therefore, for unit-normalized height interval z, the rainfall rate R_t is equal to R_{tz}/z cm/s.

$$R_t = (R_{tz} / z) \times 3600 \times 10 \, \text{mm/h}.$$

In the above equation, W_T in cm/s is the terminal velocity of the raindrop of mean volume radius R_m at level z. Rainfall rate (R_t) derived above is representative of the mean cloud scale rain intensity.

Rain formation computed for two different cloud-base CCN mean volume radius, namely 2.2 and 2.5 µm are shown in Fig. 2.7. Rain with larger rain rate forms at a lower level and extends up to a higher level for the larger size (mean volume radius 2.5 µm) cloud-base CCN spectrum.

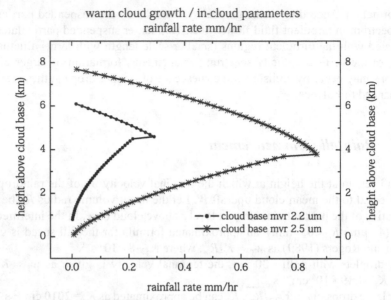

Fig. 2.7 In-cloud rain rate (mm/h)

2.3   Warm Cloud Responses to Hygroscopic Particle Seeding

In Sects. 1.3 and 1.4, it is shown that the production of buoyant energy released by condensation in turbulent eddies is mainly responsible for the formation and growth of the cloud. When warm clouds are seeded with hygroscopic particles, the turbulent buoyant energy production increases due to condensation and results in enhancement of vertical mass exchange. This would result in enhancement of convergence in the sub-cloud layer and invigoration of the updraft in the cloud. If sufficient moisture is available in the sub-cloud air layer, the enhanced convergence would lead to increased condensation and cloud growth. According to the above physical concept and theory of the cumulus cloud model presented in this chapter, it can be concluded that hygroscopic particle seeding alters the dynamics of warm clouds.

2.3.1   Dynamic Effect of Salt Seeding

Woodcock and Spencer (1967) hypothesized that dispersion of NaCl particles in a nearly water-saturated atmosphere would be sufficient to initiate a cumulus cloud. Their experiments have produced visible clouds when dry NaCl particles in the size range 0.5–20 μm diameter, were released from the aircraft in the warm moist marine boundary layer near Hawaii. The temperatures in the salt-laden air were about 0.4 °C higher than that of the environment on the average. Observations of the possible dynamic effects in warm clouds by hygroscopic particle seeding have been recorded (Murty et al. 1975, 1981; Parasnis et al. 1982). The dynamic effect of salt seeding has been estimated from the following calculations that are based on the theory and the physical concept of the cumulus cloud model discussed earlier.

The latent heat released by condensation on the hygroscopic nuclei is computed for the turbulent vertical velocities of 1 m/s and the turbulent eddy radius of 50 m. The relative humidity inside the cloud at the seeding level near the cloud base is assumed to be equal to 90 %. A seeding rate of 500 g/s and a volume dispersion rate of 45,000 m^3/s are assumed. The value of the seeding rate assumed is based on that used in the aircraft warm cloud modification experiment in India (Krishna et al. 1974; Murty et al. 1975). The value of the dispersion rate used in the computation is equal to half the value assumed by Woodcock and Spencer (1967). The median volume diameter of the salt nuclei is 10 μm. One such nucleus gives rise to the formation of a water droplet of radius 9.5 μm in a time period of 1 s at the assumed relative humidity of 90 % inside the cloud (Keith and Aarons 1954). Thus, the mass of water condensed on the seed nuclei per second per cubic metre of the seeded volume can be computed. The latent heat released by this condensation will give rise to a heating rate of 0.09 cm^{-3} s^{-1} of the seeded volume. The temperature increase will result in a vertical velocity increase of about 0.3 cm s^{-2}. This increase in the vertical velocity occurs in successive 50 m lengths of the cloud along the flight path of the aircraft whose speed is 50 m s^{-1}. The mean increase in the turbulent vertical velocity along a 500 m length of the cloud can be approximately taken to be equal to one tenth of the vertical velocity in individual 50 m length of aircraft path, i.e. equal to 0.03 cm s^{-2}. Therefore, the mean increase in the vertical velocity for the total path length of 500 m, i.e. the large eddy updraft region is equal to one fourth of the mean increase in turbulent vertical velocity (Eq. 1.1). Thus, the vertical velocity perturbation in large eddy updraft region is equal to about 0.01 cm s^{-2}. This cloud scale vertical velocity perturbation due to the hygroscopic particle seeding is equal to the naturally occurring vertical velocity perturbation at the cloud-base level for the sample cloud case discussed in Sect. 2.2. Also, it was shown that the cloud-base vertical velocity production is the main driving agent for the cloud growth processes. The sample computations of the dynamic effect of salt seeding discussed above indicated that the total cloud-growth processes in the case of seeded clouds would be increased by 100 %. Evidence of such dynamic effects due to salt seeding in warm cumulus clouds (Fig. 2.8) can be seen from the observation of cloud LWC, in-cloud temperature and cloud-top growth rates made in seeded clouds (Murty et al. 1975; Murty et al. 1981, Murty et al., 2000).

Conclusions

Atmospheric flows exhibit self-similar fractal fluctuations, a signature of long-range correlations on all space–time scales. Realistic simulation and prediction of atmospheric flows require the incorporation of the physics of observed fractal

Fig. 2.8 Seeded cloud that developed rain following seeding in 20 min. Fall streak of the raining cloud is clearly seen in the photograph (Murty et al. 2000)

fluctuation characteristics in traditional meteorological theory. A general systems theory model for fractal space–time fluctuations in turbulent atmospheric flows (Selvam 1990, 2007, 2009, 2012a, b, 2013, 2014) is presented. Classical statistical physical concepts underlie the physical hypothesis relating to the dynamics of the atmospheric eddy systems and satisfy the maximum entropy principle of statistical physics. The model predictions are as follows:

- Fractal fluctuations of atmospheric flows signify an underlying eddy continuum with overall logarithmic spiral trajectory enclosing a nested continuum of vortex-roll circulations that trace the quasi-periodic Penrose-tiling pattern. Satellite images of cyclones and hurricanes show the vivid logarithmic spiral cloud pattern whose precise mathematical equiangular spiral (golden) geometry has been used by meteorologists (Senn et al. 1957; Senn and Hisar 1959; Sivaramakrishnan and Selvam 1966; Wong et al. 2008) for locating the centre (eye) of the storm.
- The logarithmic law of wall for boundary layer flows is derived as a natural consequence of eddy mixing process. The von Karman's constant is equal to $1/\tau^2$ (≈ 0.38), where τ is the golden mean (≈ 0.618).

- The probability distribution for amplitude as well as variance of fractal fluctuations of meteorological parameters are given by the same universal inverse power-law function P, namely $P \sim \tau^{-4t}$ where the normalized standard deviation t designates the eddy length step growth number. Such a result that the additive amplitudes of eddies when squared represent the probability densities of the fluctuations is observed in the subatomic dynamics of quantum systems. Therefore, fractal fluctuations are signatures of quantum-like chaos in dynamical systems.

- The probability distribution P of amplitudes of fractal fluctuations is close to the statistical normal distribution for values of normalized standard deviation t values equal to or less than two. The probability of occurrence of extreme events, i.e. normalized deviation t values greater than two, is close to zero as given by the statistical normal distribution, while P distribution for fractal fluctuations gives appreciably high values as observed in practice.

- Atmospheric eddy energy (variance) spectrum follows the universal inverse power-law form $P = \tau^{-4t}$ indicating long-range space–time correlations between local (small-scale) and global (large-scale) perturbations.

- Atmospheric particulates are suspended in the fractal fluctuations of vertical velocities with amplitudes given by the universal inverse power-law P. A universal scale-independent mass or radius size distribution for homogeneous suspended atmospheric particulates is expressed as a function of the golden mean τ, the total number concentration and the mean volume radius. The general systems theory model for aerosol size distribution is scale-free and is derived directly from atmospheric eddy dynamical concepts. At present, empirical models such as the log-normal distribution with arbitrary constants for the size distribution of atmospheric suspended particulates are used for quantitative estimation of earth-atmosphere radiation budget related to climate warming/cooling trends.

- Numerical computations of cloud parameters were performed for two different cloud-base CCN mean volume radius, namely 2.2 and 2.5 μm and computed values are compared with the observations. Cloud-base vertical velocity production by MFC is the main driving agent for the cloud growth processes. The cloud-growth time is about 30 min as observed in practice (McGraw and Liu 2003) and is the same for the two CCN spectra since the primary trigger for cloud growth is the persistent turbulent energy generation by condensation at the cloud base in primary turbulent eddy fluctuations of radius r_* and perturbation speed w_*. However, for the larger CCN mean volume radius, namely 2.5 μm, raindrops form earlier at a lower level and extend up to higher levels in the cloud. Under suitable conditions of humidity and moisture in the environment, warm rain formation can occur at a time interval short as 30 min.

- Hygroscopic particle seeding alters the dynamics of warm clouds and enhances rainfall up to 100 % under favourable conditions of moisture supply in the environment.

References

Keith CH, Arons AB (1954) The growth of sea-salt particles by condensation of atmospheric water vapour. J Meteor 11:173–184

Krishna K, Murty ASR, Kapoor RK, Ramanamurty BhV (1974) Results of warm cloud seeding experiments in three different regions in India during the summer monsoon of 1973. Proc. IV Conf. Wea. Modification 18-21 November 1974, Fort Lauderdale, Florida, Amer. Meteor. Soc. 79–84

McGraw R, Liu Y (2003) Kinetic potential and barrier crossing: a model for warm cloud drizzle formation. Phys Rev Lett 90(1):018501–018501

Murty ASR, Selvam AM, Ramanamurty BhV (1975) Summary of observations indicating dynamic effect of salt seeding in warm cumulus clouds. J Appl Meteor 14:629–637

Murty ASR, Selvam AM, Bandyopadhyay BK, Revathy N, Pillai AG, Ramanamurty BhV (1981) Electrical and microphysical responses to salt seeding in maritime cumulus clouds. J Wea Modification 13:174–176

Murty ASR, Selvam AM, Devara PCS, Krishna K, Chatterjee RN, Mukherjee BK, Khemani LT, Momin GA, Reddy RS, Sharma SK, Jadhav DB, Vijayakumar R, Raj PE, Manohar GK, Kandalgaonkar SS, Paul SK, Pillai AG, Parasnis SS, Kulkarni CP, Londhe AL, Bhosale CS, Morwal SB, Safai PD, Pathan JM, Indira K, Naik MS, Rao PSP, Sikka P, Dani KK, Kulkarni MK, Trimbake HK, Sharma PN, Kapoor RK, Tinmaker MIR (2000) 11-year warm cloud modification experiment in Maharashtra State, India. J Wea Modification 32:10–20

Pandithurai G, Dipu S, Prabha TV, Maheskumar RS, Kulkarni JR, Goswami BN (2012) Aerosol effect on droplet spectral dispersion in warm continental cumuli. J Geophys Res 117:D16202

Parasnis SS, Selvam AM, Murty ASR, Ramanamurty BhV (1982) Dynamic responses of warm monsoon clouds to salt seeding. J Wea Modification 14:35–37

Pruppacher HR, Klett JD (1997) Microphysics of clouds and precipitation. Kluwer Academic, The Netherlands

Selvam AM, Murty ASR, Vijayakumar R, Ramanamurty BhV (1976) Aircraft measurement of electrical parameters inside monsoon clouds. Indian J Meteor Hydrol Geophys 27:391–396

Selvam AM, Murty ASR, Vijayakumar R, Paul SK, Manohar GK, Reddy RS, Mukherjee BK, Ramanamurty BhV (1980) Some thermodynamical and microphysical aspects of monsoon clouds. Proc Indian Acad Sci 89A:215–230

Selvam AM, Murty ASR, Ramanamurty BhV (1982a) Dynamics of the summer monsoon warm clouds. Proc. Reg. Sci. Conf. Trop. Meteor., 18–22 Oct. 1982, Tsukaba, Japan, 247–248

Selvam AM, Parasnis SS, Murty ASR, Ramanamurty BhV (1982b) Evidence for cloud-top entrainment in the summer monsoon warm cumulus clouds. Proc. Conf. Cloud Phys. 15–18 Nov. 1982, Chicago, Illinois, Amer. Meteor. Soc. 151–154

Selvam AM, Sikka P, Vernekar KG, Manohar GK, Mohan B, Kandalgaonkar SS, Murty ASR, Ramanamurty BhV (1982c) Temperature and humidity spectra in cloud and cloud-free air and associated cloud electrical and microphysical characteristics. Proc. Conf. Cloud Phys. 15–18 Nov. 1982, Chicago, Illinois, Amer. Meteor. Soc. 32–35

Selvam AM, Londhe AL, Vernekar KG, Mohan B, Murty ASR, Ramanamurty BhV (1982d) Aircraft observations of turbulent fluxes of momentum, heat and moisture in the sub-cloud layer and associated cloud microphysical and electrical characteristics. Proc. Conf. Cloud Phys. 15–18 Nov. 1982, Chicago, Illinois, Amer. Meteor. Soc. 42–44

Selvam AM, Murty ASR, Ramanamurty BhV (1983) Surface frictional turbulence as an agent for the maintenance and growth of large eddies in the atmospheric planetary boundary layer. Proc. VI Symp. Turbul. & Diffus., 22–25 March 1983, Boston, Mass., Amer. Meteor. Soc., 106–109

Selvam AM, Murty ASR, Ramanamurty BhV (1984a) A new hypothesis for vertical mixing in clouds. Preprint volume, 9th Intl. Cloud Phys. Conf. Tallinn, USSR

Selvam AM, Murty ASR, Ramanamurty BhV (1984b) Role of frictional turbulence in the evolution of cloud systems. Preprint volume, 9th Intl. Cloud Phys. Conf., Tallinn, USSR

Selvam AM (1990) Deterministic chaos, fractals and quantumlike mechanics in atmospheric flows. Can J Phys 68, 831–841 (http://xxx.lanl.gov/html/physics/0010046)

Selvam AM, Vijayakumar R, Murty ASR (1991a) Some physical aspects of summer monsoon clouds- comparison of cloud model results with observations. Adv Atmos Sci 8(1):111–124

Selvam AM, Vijayakumar R, Manohar GK, Murty ASR (1991b) Electrical, microphysical and dynamical observations in summer monsoon clouds. Atmos Res 26:19–32

Selvam AM (2007) Chaotic climate dynamics. Luniver, UK

Selvam AM (2009) Fractal fluctuations and statistical normal distribution. Fractals 17(3), 333–349 (http://arxiv.org/pdf/0805.3426)

Selvam AM (2012a) Universal spectrum for atmospheric suspended particulates: comparison with observations. Chaos Complex Lett 6(3), 1–43 (http://arxiv.org/abs/1005.1336)

Selvam AM (2012b) Universal spectrum for atmospheric aerosol size distribution: comparison with PCASP-B observations of vocals 2008. Nonlinear Dyn Syst Theory 12(4), 397–434 (http://arxiv.org/abs/1105.0172)

Selvam AM (2013) Scale-free Universal Spectrum for Atmospheric Aerosol Size Distribution for Davos, Mauna Loa and Izana. Int J Bifurcation Chaos 23:1350028

Selvam AM (2014) A general systems theory for rain formation in warm clouds. Chaos Complex Lett 8(1), 1–42 (http://arxiv.org/pdf/arxiv1211.0959)

Senn HV, Hiser HW, Bourret RC (1957) Studies of hurricane spiral bands as observed on radar. Final Report, U. S. Weather Bureau Contract No. Cwb-9066, University of Miami, 21pp. [NTIS-PB-168367]

Senn HV, Hiser HW (1959) On the origin of hurricane spiral bands. J Meteor 16:419–426

Sivaramakrishnan MV, Selvam MM (1966) On the use of the spiral overlay technique for estimating the center positions of tropical cyclones from satellite photographs taken over the Indian region. Proc. 12th conf. Radar Meteor., 440-446

Warner J (1970) The micro structure of cumulus clouds Part III: the nature of updraft. J Atmos Sci 27:682–688

Winkler P, Junge CE (1971) Comments on "Anomalous deliquescence of sea spray aerosols". J Appl Meteor 10:159–163

Winkler P, Junge C (1972) The growth of atmospheric aerosol particles as a function of the relative humidity. J de Recherches Atmospheriques 6:617–637

Wong KY, Yip CL, Li PW (2008) Automatic identification of weather systems from numerical weather prediction data using genetic algorithm. Expert Syst Appl 35(1–2):542–555

Woodcock AH, Spencer AT (1967) Latent heat released experimentally by adding Sodium Chloride particles to the atmosphere. J Appl Meteor 6:95–101

Chapter 3
Universal Spectrum for Atmospheric Suspended Particulates: Comparison with Observations: Data Set I

Abstract Atmospheric flows exhibit self-similar fractal space–time fluctuations on all space–time scales in association with inverse power law distribution for power spectra of meteorological parameters such as wind, temperature, etc., and thus implies long-range correlations, identified as self-organized criticality generic to dynamical systems in nature. A general systems theory based on classical statistical physical concepts discussed in Chaps. 1 and 2 visualizes the fractal fluctuations to result from the coexistence of eddy fluctuations in an eddy continuum, the larger scale eddies being the integrated mean of enclosed smaller scale eddies. The model satisfies the maximum entropy principle and predicts that the probability distributions of component eddy amplitudes and the corresponding variances (power spectra) are quantified by the same universal inverse power law distribution which is a function of the golden mean. Atmospheric particulates are held in suspension by the vertical velocity distribution (spectrum). The atmospheric particulate size spectrum is derived in terms of the model predicted universal inverse power law characterizing atmospheric eddy spectrum. Model predicted spectrum is in agreement with the four experimentally determined data sets: (i) CIRPAS mission TARFOX_WAL-LOPS_SMPS aerosol size distributions, (ii) CIRPAS mission ARM-IOP (Ponca City, OK) aerosol size distributions, (iii) SAFARI 2000 CV-580 (CARG Aerosol and Cloud Data) cloud drop size distributions (DSDs), and (iv) TWP-ICE (Darwin, Australia) rain DSDs.

Keywords Universal spectrum for atmospheric suspended particulates · Fractal fluctuations in atmospheric flows · Chaos and nonlinear dynamics · TARFOX and ARM-IOP aerosol size spectra · SAFARI 2000 cloud drop size spectra · TWP-ICE (Darwin, Australia) rain drop size spectra

3.1 Introduction

Information on the size distribution of atmospheric suspended particulates (aerosols, cloud drops, and raindrops) is important for the understanding of the physical processes relating to the studies in weather, climate, atmospheric electricity, air pollution, and aerosol physics. Atmospheric suspended particulates affect the radiative balance of the Earth/atmosphere system via the direct effect whereby they scatter

© The Author(s) 2015   55
A. M. Selvam, *Rain Formation in Warm Clouds*, SpringerBriefs in Meteorology,
DOI 10.1007/978-3-319-13269-3_3

and absorb solar and terrestrial radiation, and via the indirect effect whereby they modify the microphysical properties of clouds thereby affecting the radiative properties and lifetime of clouds (Haywood et al. 2003). At present empirical models for the size distribution of atmospheric suspended particulates are used for quantitative estimation of earth-atmosphere radiation budget related to climate warming/cooling trends. The empirical models for different locations at different atmospheric conditions, however, exhibit similarity in shape implying a common universal physical mechanism governing the organization of the shape of the size spectrum. The pioneering studies during the last three decades by Lovejoy and his group (Lovejoy and Schertzer 2008, 2010) show that the particulates are held in suspension in turbulent atmospheric flows which exhibit self-similar fractal fluctuations on all scales ranging from turbulence (mm-sec) to climate (kms-years). Lovejoy and Schertzer (2008) have shown that the rain drop size distribution (DSD) should show a universal scale invariant shape. In this study, a general systems theory for fractal space—time fluctuations developed by the author (Selvam 1990, 2005, 2007, 2009) (see Sect. 1.3) is applied to derive a universal (scale independent) spectrum for suspended atmospheric particulate size distribution expressed as a function of the golden mean τ (≈ 1.618), the total number concentration and the mean volume radius (or diameter) of the particulate size spectrum. A knowledge of the mean volume radius and total number concentration is sufficient to compute the total particulate size spectrum at any location. Model predicted spectrum is in agreement with the following four experimentally determined data sets: (i) CIRPAS mission TARFOX_WALLOPS_SMPS aerosol size distributions, (ii) CIRPAS mission ARM-IOP (Ponca City, OK) aerosol size distributions, (iii) SAFARI 2000 CV-580 (CARG Aerosol and Cloud Data) cloud DSDs, and (iv) TWP-ICE (Darwin, Australia) rain DSDs (Selvam 2012).

3.2 Atmospheric Suspended Particulates: Current State of Knowledge

3.2.1 Aerosol Size Distribution

As aerosol size is one of the most important parameters in describing aerosol properties and their interaction with the atmosphere, its determination and use is of fundamental importance. Aerosol size covers several decades in diameter and hence a variety of instruments are required for its determination. This necessitates several definitions of the diameter, the most common being the geometric diameter d. The size fraction with $d > 1–2$ μm is usually referred to as the coarse mode, and the fraction $d < 1–2$ μm is the fine mode. The latter mode can be further divided into the accumulation $d \sim 0.1–1$ μm, Aitken $d \sim 0.01–0.1$ μm, and nucleation $d < 0.01$ μm modes. Due to the d^3 dependence of aerosol volume (and mass), the coarse mode is typified by a maximum volume concentration and, similarly, the accumulation mode by the surface area concentration and the Aitken and nucleation modes by the number concentration.

Aerosol formation arises from heterogeneous or homogeneous nucleation. The former refers to condensation growth on existing nuclei, and the latter to the formation of new nuclei through condensation. Heterogeneous nucleation occurs preferentially on existing nuclei. Condensation onto a host surface occurs at a critical supersaturation, which is substantially lower (<1–2%) than for homogeneous nucleation in the absence of impurities ($>300\%$). Examples of gas-to-particle conversion are the combustion processes and the ambient formation of nuclei from gaseous organic emissions. Particles in the Aitken/accumulation mode typically arise from either: (i) the condensation of low volatility vapors; or (ii) coagulation. Particles in the accumulation mode have a longer atmospheric lifetime than other modes, as there is a minimum efficiency in sink processes. Particles in the coarse mode are usually produced by weathering and wind erosion processes. Dry deposition (primarily sedimentation) is the primary removal process. As the sources and sinks of the coarse and fine modes are different, there is only a weak association of particles in both modes (Hewitt and Jackson 2003). The aerosol chemistry data organized first by Peter Mueller and subsequently analyzed by Friedlander and coworkers showed that the fine and coarse mass modes were chemically distinctly different (Husar 2005).

Husar (2005) has summarized the history of aerosol science as follows. The modern science of atmospheric aerosols began with the pioneering work of Christian Junge who performed the first comprehensive measurements of the size distribution and chemical composition of atmospheric aerosols (Junge 1952, 1953, 1955, 1963). Based on tedious and careful size distribution measurements performed over many different parts of the world, Junge and coworkers have observed that there is a remarkable similarity in the gathered size distributions (number concentration N versus radius r_a): they follow a power law function over a wide range from 0.1 to over 20 µm in particle radius.

$$\frac{dN}{d \log r_a} = c r^{-\alpha}.$$

The inverse power law exponent α of the number distribution function ranged between 3 and 5 with a typical value of 4. This power law form of the size distribution became known as the *Junge distribution* of atmospheric aerosols. In the 1960s, the physical mechanisms that were responsible for producing these similarities in the atmospheric aerosol size spectra were not known, although it was clear that homogeneous and heterogeneous nucleation, coagulation, sedimentation, and other removal processes were all influential mechanisms. In particular, it was unclear which combination of these mechanisms is responsible for maintaining the observed *quasistationary size distribution* of the size spectra.

Whitby (1973) introduced the concept of the multimodal nature of atmospheric aerosol and Jaenicke and Davies (1976)added the mathematical formalism used today. Around 1970–1971, Whitby et al. (1972) collected and analyzed several size distribution data sets arising from different locations, times, and sampling methods, and the broad range of data provided strong evidence that bimodal distribution occurs as a ubiquitous feature of atmospheric aerosols in general, though the causal processes and mechanisms were unclear. Semiquantitative explanation of the observed fine particle dynamics provided the scientific support for the bimodal concept and became the basis of regional dynamically coupled gas-aerosol models. As pointed out by Whitby (1978) and Junge (1963)an actual size distribution comes from the sum of single modes. There is an equivalency between the optical properties of a combination of several modes and a representative single mode. From previous work, it can reasonably be assumed that aerosol size distributions follow a lognormal distribution (Tanre et al. 1996). Physical size distributions can be characterized well by a trimodal model consisting of three additive lognormal distributions (Whitey 2007). Typically, the planetary boundary layer (PBL) aerosol is combination of three modes corresponding to Aitken nuclei, accumulation mode aerosols, and coarse aerosols, the shape of which is often modeled as the sum of lognormal modes (Whitey 2007; Chen et al. 2009). In a nutshell, the bimodal distribution concept states that the atmospheric aerosol mass is distributed in two distinct size ranges, fine and coarse and that each aerosol mode has a characteristic size distribution, chemical composition, and optical properties (Husar 2005).

3.3   Cloud Drop Size Distribution

3.3.1   Cloud Microphysics and Associated Cloud Dynamical Processes

Prupaccher and Klett (1997) have summarized the current state of knowledge of cloud microphysical processes as follows. One principal continuing difficulty is that of incorporating, in a physically realistic manner, the microphysical phenomena in the broader context of the highly complex macrophysical environment of natural clouds. Mason (Mason 1957) also refers to the problem of scale in cloud microphysics. Cloud microphysics deals with the growth of particles ranging from the characteristic sizes of condensation nuclei ($\leq 10^{-2}$ μm) to precipitation particles ($\leq 10^4$ μm for raindrops, $\leq 10^5$ μm for hailstones). This means we must follow the evolution of the particle size spectrum, and the attendant microphysical processes of mass transfer, over about seven orders of magnitude in particle size. Similarly, the range of relevant cloud–air motions varies from the characteristic size of turbulent eddies which are small enough to decay directly through viscous dissipation ($\leq 10^{-2}$ cm), since it is these eddies which turn out to define the characteristic shearing rates for turbulent aerosol coagulation processes, to motion on scales at least as large as the cloud itself ($> 10^5$ cm). Thus, relevant interactions may occur over at least seven orders of magnitude of eddy sizes. A complete in-context understanding of cloud microphysics including dynamic, electrical, and chemical effects is not yet available. Many microphysical mechanisms are still not understood in quantitative detail (Pruppacher and Klett 1997).

Although the relative humidity of clouds and fogs usually remains close to 100%, considerable departures from this value have been observed. The spatial and temporal nonuniformity of the humidity inside clouds and fogs results in a corresponding rapid spatial variation of the concentration of cloud drops and the cloud liquid water content. Based on his observations, Warner (1969) suggested that bimodal DSDs are the result of a mixing process between the cloud (cumulus) and the environment. Warner proposed that the mixing process producing the bimodality is due mostly to entrainment of drier air at the growing cloud top, and to a lesser degree, to entrainment at the cloud edges. The size distribution experiences a broadening effect with increasing distance from cloud base. Spectra with double maxima have also been observed by others in other regions. If we consider the spatial distribution of the drop size, number concentration, and liquid water content, we find strongly inhomogeneous conditions. The cloud liquid water content w_L varies rapidly over short distances along a horizontal flight path in a manner which is closely related to the variation of the vertical velocity in the cloud and also w_L varies essentially as the total number concentration of drops. Vulfson et al. (1973) demonstrate that the cloud water content typically increases with height above the cloud base, assumes a maximum somewhere in the upper half of the cloud, and then decreases again toward the cloud top. In most cases, a comparison between the observed cloud water content w_L and adiabatic liquid water content $(w_L)_{ad}$ computed on the basis of a saturated adiabatic ascent of moist air shows that generally $w_L < (w_L)_{ad}$. In most cases, $w_L/(w_L)_{ad}$ is found to decrease with increasing height above cloud base but to increase with cloud width. This implies that the entrainment is especially pronounced near the cloud top, while the net dilution effect by entrainment is less in wider clouds than narrower ones.

3.3.2 Formulations for DSDs in Clouds and Fog

For many fog and cloud modeling purposes, it is necessary to be able to approximate the observed DSD by an analytical expression. Fortunately, DSDs measured in many different types of clouds and fogs under a variety of meteorological conditions often exhibit a characteristic shape. Generally, the concentration rises sharply from a low value to a maximum, and then decreases gently towards larger sizes. Such a characteristic shape can be approximated reasonably well by either a gamma distribution or a lognormal distribution. In order to describe a DSD with two or more maxima, one or more unimodal distributions may be superposed. As an example, according to Khrgian and Mazin (1952)(in Borovikov et al. (1963), many DSDs with a single maximum may also be quite well be represented by a gamma distribution. Another convenient representation of the cloud DSD is the empirical formula developed by Best (1951a, 1951b). These various analytical expressions only represent average distributions. Individual drop size spectra may be significantly different (Pruppacher and Klett 1997). A wealth of aircraft measurements in the Soviet Union indicates that droplet size spectra in stratocumulus are distributed in logarithmic normal (Levin 1954) or gamma (Borovikov et al. 1963) forms. The droplet size spectra in stratus and stratocumulus are now commonly described by the gamma distribution. Droplet size spectra in altostratus and altocumulus as a function of temperature and cloud thickness are given according to Mazin and Khrgian (1989) (Hobbs 1993).

3.4 Rain Drop Size Distribution

3.4.1 Classical Cloud Microphysical Concepts

Rain drops are large enough to have a size-dependent shape which cannot be characterized by a single length. The conventional resolution is to describe rain spectra in terms of the equivalent diameter D_0 defined as the diameter of a sphere of the same volume as the deformed drop. The overall shaping of the spectrum is obviously quite complicated, and determined in part by such meteorological variables as temperature, relative humidity, and wind in the subcloud region.

Various empirical relations have been advanced to describe the size spectra of raindrops. One often used is the size distribution proposed by Best (1950). Probably the most widely used description for the raindrop spectrum is the size distribution of Marshall and Palmer (MP) (Marshall and Palmer 1948), which is based on the observations of Laws and Parsons (1943). More detailed studies have demonstrated that the MP distribution is not sufficiently general to describe most observed raindrop spectra accurately.

Numerous studies have also used the gamma distribution. Another alternative is the log normal distribution. Detailed comparison between the raindrop spectra actually observed and these empirical distributions show that in most cases only a partial fit can be achieved at best.

The observed raindrop spectra also show, apart from a main mode, some secondary modes. It is reasonable to attribute the main mode as well as the subpeaks to collisional drop breakup. Unexpectedly, these peaks are not present in all rain DSDs. One explanation for this may be that the breakup-induced peaks become masked due to turbulent and evaporative effects. Additional factors which complicate an interpretation of observed raindrop distributions are related to instrumental problems (Pruppacher and Klett 1997).

3.4.2 Cloud Microphysics and Self-Similar Turbulent Atmospheric Flows

Lovejoy and his group (Lovejoy and Schertzer 2008, 2010) have done pioneering studies on self-similar fractal fluctuations ubiquitous to turbulent atmospheric flows and have emphasized the urgent need to incorporate, in modeling studies of microphysics of clouds and rain, the theory of nonlinear dynamical systems as summarized in the following.

Rain is a highly turbulent process yet there is a wide gap between the turbulence and precipitation research. It is still common for turbulence to be invoked as a source of homogenization, an argument used to justify the use of homogeneous (white noise) Poisson process models of rain.

Dimensional analysis shows that the cumulative probability distribution of nondimensional drop mass should be a universal function dependent only on scale. Starting in the 1980s, a growing body of the literature has demonstrated that—at least over large enough scales involving large numbers of drops—rain has nontrivial space–time scaling properties. While the traditional approach to drop modeling is to hypothesize specific parametric forms for the DSD and then to assume spatial homogeneity in the horizontal and smooth variations in the vertical, the nonlinear dynamics approach on the contrary assumes extreme turbulent-induced variability governed by the turbulent cascade processes and allows the DSD to be constrained by the turbulent fields.

The conventional methods of modeling the evolution of raindrops give turbulence at most a minor (highly "parameterized") role: the atmosphere is considered homogeneous and the spatial variability of the DSD arises primarily due to complex drop interactions.

It is shown on dimensional grounds that the dimensionless cumulative DSD as a function of the dimensionless drop mass should be a universal function of dimensionless mass (Lovejoy and Schertzer 2008). Khain et al. (2007) have given critical comments to results of investigations of drop collisions in turbulent clouds and conclude that the fact that turbulence enhances the rate of particle collisions can be considered as being established.

3.5   Data

Four data sets, namely, two aerosol (I and II), one cloud drop size (III), and one rain drop size (IV) were used for comparison of observed with model predicted suspended particle size spectrum in turbulent atmospheric flows.

3.5.1   Data Set I, Aerosol Size Spectrum

TARFOX_WALLOPS_SMPS: Tropospheric Aerosol Radiative Forcing Observational eXperiment (TARFOX), Langley DAAC Project—Scanning Mobility Particle Sizer (TSI Incorporated, St. Paul, MN) data taken at Wallops ground station (37.85° lat, −75.48° lon) in the US Eastern seaboard. Ground-based ambient size distribution of aerosol (10.7–749 nm diameter, fine mode) at point measurements at 5 min time intervals was taken during the period 10th–31st July 1996. Raw data exported using SMPS 2.0 (TSI), then imported to Microsoft Excel, adjusted to local time, then saved as comma delimited. Data were obtained from http://eosweb.larc. nasa.gov/PRODOCS/tarfox/table_tarfox.html and http://eosweb.larc.nasa.gov/cgi-bin/searchTool.cgi?Dataset=TARFOX_WALLOPS_SMPS

3.5.2   Data Set II, Aerosol Size Spectrum

PCASP files (replaced on January 14, 2005): Contain 1 Hz size distribution data measured aboard the Center for Interdisciplinary Remotely-Piloted Aircraft Studies (CIRPAS) Twin Otter during the Atmospheric Radiation Program (ARM) Intensive Operational Period (IOP) 2003 using the PCASP, with SPP-200 electronics. PCASP is Passive Cavity Aerosol Spectrometer manufactured by PMS Inc., but with an SPP-200 data system manufactured by DMT Inc., Ponca City, Oklahoma, USA, was the base from which the flights were conducted. These were typically between 3 and 5 h flights, carried out during the month of May 2003. The Aerosol IOP was conducted between May 5 and 31, 2003 over the ARM Southern Great Plains (SGP) site. There were a total of 16 science flights, for a total of 60.6 flight hours, conducted by the CIRPAS Twin Otter aircraft on 15 days during this period. Most of the Twin Otter flights were conducted under clear or partly cloudy skies to assess aerosol impacts on solar radiation.

The aerosol particle size concentrations ranging in diameter from 0.1 to 3.169701 μm (accumulation and coarse modes) were measured in 20 channels (size ranges). The geometric mean radius of the class interval was used for computing $d(\ln r_{an})$. The data sets were obtained from ARM IOP Data Archive http://www.archive.arm.gov/armlogin/login.jsp.

3.5.3   Data Set III, Cloud Drop Size/Number Concentration

Cloud drop size/number concentration. Project SAFARI 2000, CARG Aerosol and Cloud Data from the Convair-580. Web Site: http://cargsun2.atmos.washington. edu/. The Cloud and Aerosol Research Group (CARG) of the University of Washington participated in the SAFARI-2000 Dry Season Aircraft campaign with their Convair-580 research aircraft. This campaign covered five countries in southern Africa from 10 August through 18 September, 2000 on the 31 research flights.   (http://daac.ornl.gov/data/safari2k/atmospheric/CV-580/comp/SAFARI-MASTER.pdf).

Data Citation: Hobbs P. V. 2004. SAFARI 2000 CV-580 Aerosol and Cloud Data, Dry Season 2000 (CARG). Data set. Available online (http://www.daac.ornl. gov) from Oak Ridge National Laboratory Distributed Active Archive Center, Oak Ridge, Tennessee, USA, doi:10.3334/ORNLDAAC/710. All data taken at latitude: 14.00 to 26.00 S, longitude: 36.00 to 11.00E.

Data details: cloud particle concentration per cc between 1.7 and 47.0 μm in 15 channels. Particle Measuring Systems Model FSSP-100. Calculated from raw counts and sample time. Seven data sets containing cloud drop size/number concentration were used for the study. The data are output at 1-s resolution (http://daac.ornl.gov/S2K/guides/s2k_CV580.html).

3.5.4 Data Set IV: TWP-ICE, Joss–Waldvogel Disdrometer Rain DSDs

The Tropical Western Pacific—International Cloud Experiment (TWP-ICE) was held near Darwin, Australia, to collect in-situ and remote-sensing measurements of clouds, precipitation, and meteorological variables from the ground to the lower stratosphere. During TWP-ICE, vertically pointing profiling radar, surface rain gauge, and disdrometer observations were collected for the whole wet season from November 2005 through March 2006. The Joss Waldvogel disdrometer was operational from 3 November 2005 through 10 February 2006.

3.6 Analysis and Discussion of Results

The atmospheric suspended particulate size spectrum is closely related to the vertical velocity spectrum (Sect. 1.6). The mean volume radius of suspended aerosol particulates increases with height (or reference level z) in association with decrease in number concentration. At any height (or reference level) z, the fractal fluctuations (of wind, temperature, etc.) carry the signatures of eddy fluctuations of all size scales since the eddy of length scale z encloses smaller scale eddies and at the same time forms part of internal circulations of eddies larger than length scale z (Sect. 1.4). The observed atmospheric suspended particulate size spectrum also exhibits a decrease in number concentration with increase in particulate radius. At any reference level z of measurement the mean volume radius r_{as} will serve to calculate the normalized radius r_{an} for the different radius class intervals as explained below.

The general systems theory for fractal space–time fluctuations in dynamical systems predicts universal mass size spectrum for atmospheric suspended particulates (Sect. 1.6.4). For homogeneous atmospheric suspended particulates, i.e., with the same particulate substance density, the atmospheric suspended particulate mass and radius size spectrum is the same and is given as (Sect. 1.6.4) the normalized aerosol number concentration equal to $\dfrac{1}{N}\dfrac{dN}{d(\ln r_{an})}$ versus the normalized aerosol radius r_{an}, where (i) r_{an} is equal to $\dfrac{r_a}{r_{as}}$, r_a being the mean class interval radius and r_{as} the mean volume radius for the total aerosol size spectrum; (ii) N is the total aerosol number concentration and dN is the aerosol number concentration in the aerosol radius class interval dr_a; and (iii) $d(\ln r_{an})$ is equal to $\dfrac{dr_a}{r_a}$ for the aerosol radius class interval r_a to $r_a + dr_a$.

3.6.1 Analysis Results, Data I: TARFOX_WALLOPS_SMPS, Aerosol Size Spectra

A total of 23 data sets between 16 July and 26 July 1996 are available for the study. The data consists of particle number concentration per cc in 59 class intervals ranging from 10.7 to 749 nm (fine mode) for the particle diameter. The midpoint diameter of the class interval was used to compute the corresponding value of $d(\ln r_{an})$. The average aerosol size spectra for each of the 23 data sets are plotted on the left-hand side and the total average spectrum for the 23 data sets is plotted on the right-hand side in Fig. 3.1a along with the model predicted scale independent aerosol size spectrum. The corresponding standard deviations for the average spectra are shown as error bars in Figs. 3.1a. The average values of mean volume radius (nm), total number concentration (cm^{-3}), the number of spectra, the number of particle size class intervals for each of the 23 data sets and the upper and lower bounds of particle size (diameter in nm) intervals are given in Fig. 3.1b.

The aerosol size spectra cover the fine size range and display large deviations from the mean, particularly for the individual flights (left-hand side of Fig. 3.1a). The source of the uncertainties displayed by the error bars (Fig. 3.1a) may be due to measurement noise, independent in every size interval, also may be due to different aerosol sources with different particle substance densities. However, a major portion of the total average aerosol size spectrum (right-hand side of Fig. 3.1a) shows a reasonably good fit (within plus or minus two standard deviations) to the model predicted universal spectrum for homogeneous aerosols, i.e., same aerosol source.

3.6.2 Analysis Results, Data II: CIRPAS Twin Otter Flight Data Sets, Aerosol Size Spectra

ARM Aerosol IOP at the SGP site over a 3–4 week period centered on May 2003. CIRPAS Twin Otter flight data sets using the PCASP. A total of 16 data sets are available for the study.

The aerosol particle size concentrations ranging in diameter from 0.1 to 3.169701 μm consisting of accumulation mode (up to 1 μm) and coarse mode (diameter >1 μm) were measured in 20 channels (size ranges). The geometric mean radius of the class interval was used for computing $d(\ln r_{an})$. The data sets were obtained from ARM IOP Data Archive http://www.archive.arm.gov/armlogin/login.jsp. The average aerosol size spectra for each of the 16 data sets are plotted on the left-hand side and the total average spectrum for the 16 data sets is plotted on the right and side in Fig. 3.2a along with the model predicted scale independent aerosol

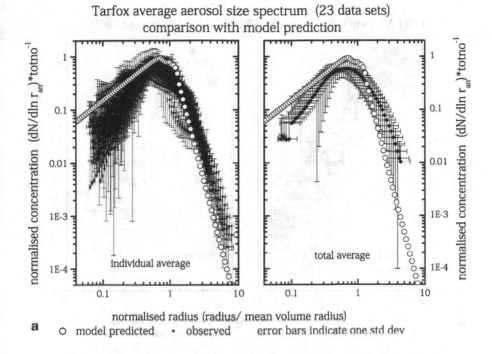

a   ○   model predicted   •   observed   error bars indicate one std dev

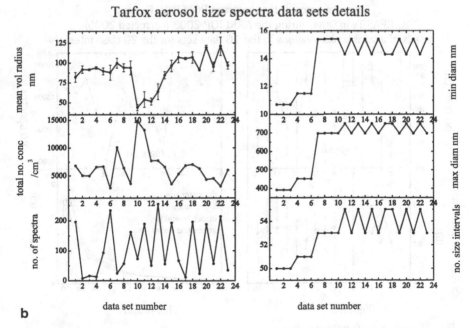

b   data set number   data set number

Fig. 3.1 a Average aerosol size spectrum for each of the 23 data sets (*left*) and total average aerosol size spectrum for the 23 data sets (*right*). Error bars indicate one standard deviation on either side of the mean. Model predicted scale independent aerosol size spectrum also is shown in the figure. **b** Details of (i) mean volume radius nm, (ii) aerosol total number concentration cm^{-3} number of spectra, (iv) lower bound, (v)upper bound diameter nm for the particle size class intervals used (total available = 59), and (vi) number of particle size class intervals for each of the 23 data sets available for TARFOX aerosol size spectra

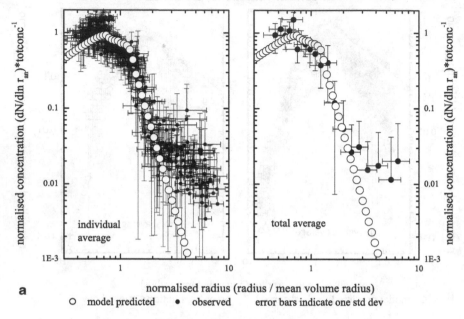

average aerosol size spectra PCASP 16 flights
CIRPAS Twin Otter during the ARM IOP 2003 (corrected 2005)

a

normalised radius (radius / mean volume radius)

○   model predicted   ●   observed   error bars indicate one std dev

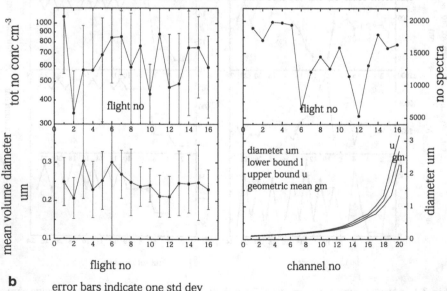

average aerosol size spectrum PCASP 16 flights
CIRPAS Twin Otter during the ARM IOP 2003 (corrected 2005)
details of available observations in the 16 data sets for the 20 class intervals

b   error bars indicate one std dev

Fig. 3.2 a Average aerosol size spectrum for each of the 16 data sets (*left*) and total average aerosol size spectrum for the 16 data sets (*right*). *Error bars* indicate one standard deviation on either side of the mean. Model predicted scale independent aerosol size spectrum also is shown in the figure. **b** Details of (i) mean volume diameter μm, (ii) aerosol total number concentration cm^{-3}, and (iii) number of spectra for the 16 data sets and the lower bound, upper bound and geometric mean diameter μm for the 20 particle size class intervals

size spectrum. The corresponding standard deviations for the average spectra are shown as error bars in Figs. 3.2a. The average values of mean volume radius, total number concentration, the number of spectra and the upper and lower bounds of particle size (radius) intervals are given in Fig. 3.2b.

The portion of the total average aerosol size spectrum corresponding to the accumulation mode (up to 1 μm or normalized diameter equal to 4) shows a reasonably good fit (within plus or minus two standard deviations) to the model predicted universal spectrum. The source of the uncertainties displayed by the error bars may be due to measurement noise, independent in every size interval, also may be due to different aerosol sources. The coarse mode (diameter > 1 μm) portion of the total average aerosol size spectrum shows significant departure from the model predicted spectrum and may be attributed to a different source region for the suspended particulates with a different density. The model predicts universal spectrum for suspended aerosol mass size distribution (Sect. 1.6.4), based on the concept that the atmospheric eddies hold in suspension the aerosols and thus the mass size spectrum of the atmospheric aerosols is dependent on the vertical velocity fluctuation spectrum of the atmospheric eddies.

3.6.3   Analysis Results, Data III: CARG Aerosol and Cloud Data from the Convair-580, Cloud Drop Size Spectra

Cloud drop size/number concentration from Project SAFARI 2000, CARG Aerosol and Cloud Data from the Convair-580. A total of 7 data sets are available for the study. Cloud drop size spectra were computed for data sets for which the cloud liquid water content (Johnson-Williams) was more than zero.

The cloud drop size/number concentrations ranging in diameter from 1.70 to 47.0 μm were measured in 15 channels (size ranges). The arithmetic mean radius of the class interval was used for computing $d(\ln r_{an})$. The data sets were obtained from ftp://ftp.daac.ornl.gov/data/safari2k/atmospheric/CV-580/data/. The average aerosol size spectra for each of the 7 data sets are plotted on the left-hand side and the total average spectrum for the 7 data sets is plotted on the right and side in Fig. 3.3a along with the model predicted scale independent aerosol size spectrum. The corresponding standard deviations for the average spectra are shown as error bars in Figs. 3.3a. The average values of mean volume radius, total number concentration, the number of spectra, the number of class intervals for each of the 7 data sets, and the upper and lower bounds of particle size (radius) intervals are given in Fig. 3.3b.

The individual and total average cloud droplet spectra show a close fit to model predicted universal spectrum for cloud drop diameter more than 5 μm (corresponding to normalized diameter equal to 1). The observed spectra show appreciably larger radii than model predicted for normalized radius size range less than 1 and may be attributed to the large increase in the sampled median volume diameter, from about 5 to 20 μm during flights 5 to 7 (Fig. 3.3b). However, even in this region for

Cloud Data from the Convair-580 during SAFARI 2000 (7 flights)
average cloud droplet size spectra / comparison with model prediction

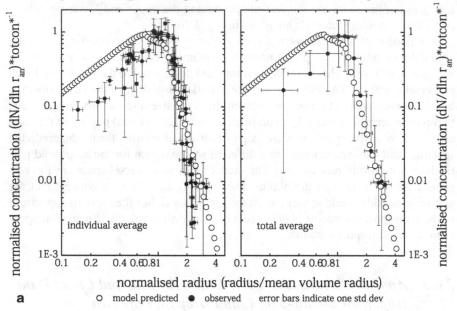

Cloud Data from the Convair-580 during SAFARI 2000
individual (7 flights) average cloud droplet size spectra
details of data sets

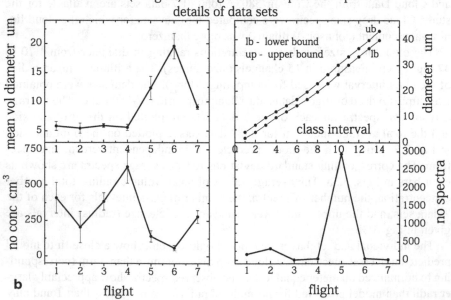

Fig. 3.3 a Average cloud drop size spectrum for each of the 7 data sets (*left*) and total average cloud drop size spectrum for the 7 data sets (*right*). *Error bars* indicate one standard deviation on either side of the mean. Model predicted scale independent suspended particulate size spectrum also is shown in the figure. **b** Details of (i) mean volume diameter μm, (ii) cloud drop total number concentration cm^{-3}, and (iii) number of spectra for each of the 7 data sets and the lower bound and upper bound diameter μm for the 15 drop size class intervals

normalized radius less than 1, the model predicted and the total average spectrum (right-hand side of Fig. 3.3a) are within two standard deviations from the mean, a standard statistical criterion for "goodness of fit."

3.6.4   Analysis Results Data IV: TWP-ICE, Joss–Waldvogel Disdrometer Rain DSDs

The Joss–Waldvogel disdrometer (http://cires.colorado.edu/blogs/twc-ice/2009/09 /19/4-0-joss-waldvogel-disdrometer) was operational from 3 November 2005 through 10 February 2006. The original JWD data were collected at the full 127 diameter channels and with a 10 s dwell time. These high-resolution data were reduced to the standard 20 diameter channels and to a 1-min resolution. A dead-time correction (Sheppard and Joe 1994; Sauvageot and Lacaux 1995) was applied to the raindrop counts. The data products are provided using the dead-time corrected raindrop counts and 60 s dwell time.

The ASCII data files are day files with 1-min resolution and contain 1440 rows. Bad or missing data values are indicated with a value of -99.9. The ASCII data files can be found on the ftp site: ftp://ftp.etl.noaa.gov/user/cwilliams/Darwin/disdrometer/dat/

The values of $d(\ln r_{an})$ for the rain drop size spectrum was calculated from the number of raindrops in each raindrop diameter size in a total of 20 standard diameter channels ranging from 0.34 to 5.37 mm and corresponding channel width (mm). A total of 99 data sets (days) are available for the study. The average (daily) rain drop size spectra for each of the 99 data sets are plotted on the left-hand side and the total average (daily) spectrum for the 99 data sets is plotted on the right-hand side in Fig. 3.4a along with the model predicted scale independent aerosol size spectrum. The corresponding standard deviations for the average spectra are shown as error bars in Fig. 3.4a. The average values of (i) mean volume diameter mm, (ii) rain drop total number, (iii) number of observations in each of the 20 channels, and (iv) channel diameter and channel width (mm) for the 20 drop size channels are given in Fig. 3.4b.

The total average raindrop size spectrum (right-hand side of Fig. 3.4a) shows a reasonably good fit (within two standard deviations from the mean) even though the individual average raindrop size spectra (left-hand side of Fig. 3.4a) show large error bars which may be attributed to the large variability in total number of drops sampled corresponding to each mean volume radius (Fig. 3.4b).

Conclusions

A general systems theory for fractal space–time fluctuations in turbulent atmospheric flows predicts a universal scale-independent mass or radius size distribution for homogeneous suspended atmospheric particulates expressed as a function of the golden mean τ, the total number concentration and the mean volume radius. Model

average raindrop size spectrum TWP-ICE Darwin, Australia
comparison with model predicted spectrum (nov 2005 - feb 2006, 99 days)

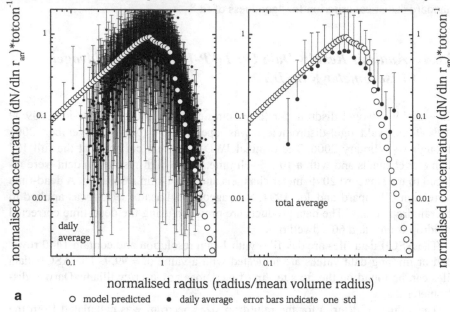

a   o   model predicted   •   daily average   error bars indicate one std

average raindrop size spectrum TWP-ICE Darwin, Australia
details of data sets (november 2005 - february 2006, 99 days)

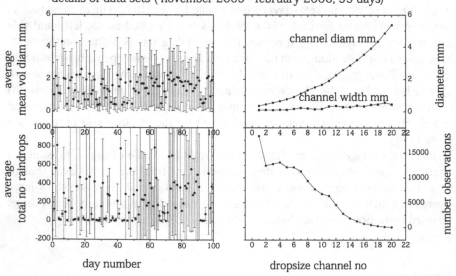

b   error bars indicate one std dev

Fig. 3.4 **a** Average (daily) rain drop size spectra for each of the 99 data sets are plotted on the left-hand side and the total average (daily) spectrum for the 99 data sets is plotted on the right-hand side along with the model predicted scale independent aerosol size spectrum. The corresponding standard deviations for the average spectra are shown as error bars. **b** Average values of (i) mean volume diameter mm, (ii) rain drop total number, (iii) number of observations in each of the 20 channels, and (iv) channel diameter and channel width (mm) for the 20 drop size channels

predicted spectrum is in agreement (within two standard deviations on either side of the mean) with total averaged radius size spectra for the following four experimentally determined data sets: (i) CIRPAS mission TARFOX_WALLOPS_SMPS aerosol size distributions; (ii) CIRPAS mission ARM-IOP (Ponca City, OK)aerosol size distributions; (iii) SAFARI 2000 CV-580 (CARG Aerosol and Cloud Data) cloud DSDs; and (iv) TWP-ICE (Darwin, Australia) rain DSDs. SAFARI 2000 aerosol size distributions reported by Haywood et al. (2003) also show similar shape for the distributions. Classical statistical physical concepts underlie the physical hypothesis relating to the dynamics of the atmospheric eddy systems proposed in this chapter. Model predicted spectrum satisfies the maximum entropy principle of statistical physics.

The general systems theory model for aerosol size distribution is scale free and is derived directly from atmospheric eddy dynamical concepts. At present empirical models such as the log normal distribution with arbitrary constants for the size distribution of atmospheric suspended particulates are used for quantitative estimation of earth-atmosphere radiation budget related to climate warming/cooling trends (Sect. 3.2).

References

Best AC (1950) The size distribution of raindrops. Quart J R Met Soc 76:16–36

Best AC (1951a) The size of cloud droplets in layer-type cloud. Quart J Met Soc 77:241-248

Best AC (1951b) Drop-size distribution in cloud and fog. Quart J R Met Soc 77:418-426

Borovikov AM, Gaivoronskii II, Zak EG, Kostarev VV, Mazin IP, Minervin VE, Khrgian AKh, Shmeter SM (1963) Cloud physics Israel program of scientific translations, Jerusalem, pp 393

Chen WN, Chang SY, Chou CCK, Fang GC (2009) Total scatter-to-backscatter ratio of aerosol derived from aerosol size distribution measurement. Int J Environ Pollut 37(1):45–54

Haywood J, Francis P, Dubovik O, Glew M, Holben B (2003) Comparison of aerosol size distributions, radiative properties, and optical depths determined by aircraft observations. J Geophy Res 108(D13):8471

Hewitt CN, Jackson AV: Handbook of atmospheric science: principles and applications. Wiley-Blackwell, USA (2003)

Hobbs PV (1993) Aerosol-cloud-climate interactions. 54 (International Geophysics). Academic, USA

Hobbs PV (2004) SAFARI 2000 CV-580 Aerosol and Cloud Data, Dry Season 2000 (CARG). Data set. Available on-line retrieved 28 December 2014 (http://www.daac.ornl.gov) from Oak Ridge National Laboratory Distributed Active Archive Center, Oak Ridge, Tennessee, U.S.A. doi:10.3334/ORNLDAAC/710

Husar RB (2005) Emergence of the bimodal distribution concept. In: Sem GJ, Boulaud D, Brimblecombe. P, Ensor ES, Gentry JW, Marijnissen JCM, Preining O (eds) History & reviews of aerosol science. American Association for Aerosol Research, Portland

Jaenicke R, Davies CN (1976) The mathematical expression of the size distribution of atmospheric aerosols. J Aerosol Sci 7:255–259

Junge C (1952) Gesetzmäßigkeiten in der Größenverteilung atmosphärischer Aerosoleüber dem Kontinent. Ber Deut Wetterdienst US-Zone 35:261–277

Junge CE (1953) Die Rolle der Aerosols und der gasformigen Beimengungen der Luft im Spurenstoffhaushalt der Troposhere. Tellus 5:1–26

Junge CE (1955) The size distribution and aging of natural aerosols as determined from electrical and optical data on the atmosphere. J Met 12:13–25

Junge CE (1963) Air chemistry and radioactivity. Academic, London, p 382

Khain A, Pinsky M, Elperin T, Kleeorin N, Rogachevskii I, Kostinski A (2007) Critical comments to results of investigations of drop collisions in turbulent clouds. Atmos Res 86:1–20

Khrgian AKh, Mazin IP (1952) The size distribution of droplets in clouds. Trudy TsAo, 7:56.

Laws JO, Parsons DA (1943) The relation of raindrop-size to intensity, Eos Trans. AGU, 24(2), 452–460

Levin LM (1954) Size distribution function for cloud droplets and raindrops. Dokl Akad Nauk SSSR 94:1045–1053

Lovejoy S, Schertzer D (2008) Turbulence, raindrops and the $l^{1/2}$ number density law. New J Phys 10(075017):1–32

Lovejoy S, Schertzer D (2010) Towards a new synthesis for atmospheric dynamics: space-time cascades. Atmos Res 96:1–52

Marshall JS, Palmer WMcK (1948) The distribution of raindrops with size. J Met 5:165–166

Mason BJ (1957) The physics of clouds. Oxford University, Oxford

Mazin IP, Khrgian AK (1989) Handbook of clouds and cloudy atmospheres. Gidrometeoizdat, Leningrad

Pruppacher HR, Klett JD (1997) Microphysics of clouds and precipitation. Kluwer Academic, The Netherlands

Sauvageot H, Lacaux J-P (1995) The shape of averaged drop size distributions. J Atmos Sci 52:1070–1073

Selvam AM (1990) Deterministic chaos, fractals and quantumlike mechanics in atmospheric flows. Can J Phys 68:831–841. http://xxx.lanl.gov/html/physics/0010046

Selvam AM (2005) A general systems theory for chaos, quantum mechanics and gravity for dynamical systems of all space-time scales. Electromagn Phenom 5 No.2(15):160–176. http://arxiv.org/pdf/physics/0503028; http://www.emph.com.ua/15/selvam.htm

Selvam AM (2007) Chaotic climate dynamics. Luniver, U. K

Selvam AM (2009) Fractal fluctuations and statistical normal distribution. Fractals 17(3):333–349. http://arxiv.org/pdf/0805.3426

Selvam AM (2012) Universal spectrum for atmospheric suspended particulates: comparison with observations. Chaos & complex. Lett 6(3):1–43. http://arxiv.org/abs/1005.1336

Sheppard BE, Joe PI (1994) Comparison of raindrop size distribution measurements by a Joss–Waldvogel disdrometer, a pms 2dg spectrometer, and a Poss Doppler radar. J Atmos Ocean Tech 11:874–887

Tanre D, Herman M, Kaufman YJ (1996) Information on aerosol size distribution contained in solar reflected spectral radiances. J Geophys Res 101(D14):19043–19060

Vulfson NI, Laktionov AG, Skatsky VI (1973) Cumuli structure of various stages of development. J Appl Met 12:664–670

Warner J (1969) The microstructure of cumulus cloud. Part I. general features of the droplet spectrum. J Atmos Sci 26:1049–1059

Whitby KT (1973) On the multimodal nature of atmospheric aerosol size distribution. Paper presented at the VIII International Conference on Nucleation, Leningrad

Whitby KT (1978) The physical characteristics of sulphur aerosols. Atmos Environ 12:135–159

Whitey KT (2007) The physical characteristics of sulfur aerosols. Atmos Environ 41(Supplement 1):25–49

Whitby KT, Husar RB, Liu BYH (1972) The aerosol size distribution of Los Angeles smog. J Colloid Interface Sci 39:177–204

Chapter 4
Universal Spectrum for Atmospheric Suspended Particulates: Comparison with Observations: Data Set II

Abstract Atmospheric flows exhibit self-similar fractal space–time fluctuations on all space–time scales in association with inverse power law distribution for power spectra of meteorological parameters such as wind, temperature, etc., and thus implies long-range correlations, identified as self-organized criticality generic to dynamical systems in nature. A general systems theory visualizes the fractal fluctuations to result from the coexistence of eddy fluctuations in an eddy continuum, the larger scale eddies being the integrated mean of enclosed smaller scale eddies. The model predicts that the probability distributions of component eddy amplitudes and the corresponding variances (power spectra) are quantified by the same universal inverse power law distribution incorporating the golden mean. Atmospheric particulates are held in suspension by the vertical velocity distribution spectrum. The atmospheric particulate size spectrum is derived in terms of the model-predicted universal inverse power law characterizing atmospheric eddy spectrum. Model-predicted spectrum is in agreement with PCASP-B aerosol-size spectra measurements made during the VOCALS ((VAMOS Ocean-Cloud-Atmosphere Land Study) 2008.

Keywords Universal spectrum for atmospheric suspended particulates · Fractal fluctuations in atmospheric flows · Chaos and nonlinear dynamics · Maximum entropy principle · VOCALS 2008 PCASP-B aerosol size spectra

4.1 Introduction

A general systems theory for fractal space–time fluctuations (Selvam 1990 2005, 2007, 2009) (see Sect. 1.3) is applied to derive universal (scale-independent) inverse power law distribution incorporating the golden mean for atmospheric eddy energy distribution. Atmospheric particulates are held in suspension by the spectrum of atmospheric eddy fluctuations (vertical). The suspended atmospheric particulate size distribution is expressed in terms of the atmospheric eddy energy spectrum and is expressed as a function of the golden mean $\tau(\approx 1.618)$, the total number concentration and the mean volume radius (or diameter) of the particulate size spectrum. A knowledge of the mean volume radius and total number concentration is sufficient to compute the total particulate size spectrum at any location. Model predicted atmospheric eddy energy spectrum is in agreement with earlier observational

© The Author(s) 2015
A. M. Selvam, *Rain Formation in Warm Clouds*, SpringerBriefs in Meteorology,
DOI 10.1007/978-3-319-13269-3_4

results (Selvam et al. 1992; Selvam and Joshi 1995; Selvam et al. 1996; Selvam and Fadnavis 1998; Joshi and Selvam 1999; Selvam 2011). Model-predicted suspended particulate (aerosol) size spectrum (Sect. 1.6.4) is in agreement with observations using VOCALS 2008 PCASP data (Selvam 2012).

4.2 Data

VOCALS PCASP-B data sets were used for comparison of observed with model-predicted suspended particle-size spectrum in turbulent atmospheric flows.

During October and November, 2008, Brookhaven National Laboratory (BNL) participated in VOCALS (VAMOS Ocean-Cloud-Atmosphere Land Study), a multiagency, multinational atmospheric sampling field campaign conducted over the Pacific Ocean off the coast of Arica, Chile. Support for BNL came from U.S. Department of Energy (DOE)'s Atmospheric Science Program (ASP) which is now part of the Atmospheric System Research (ASR) program following a merger with DOE's Atmospheric Radiation (ARM) program. A description of the VOCALS field campaign can be found at http://www.eol.ucar.edu/projects/vocals/

Measurements made from the DOE G-1 aircraft are being used to assess the effects of anthropogenic and biogenic aerosol on the microphysics of marine stratus. Aerosols affect the size and lifetime of cloud droplets thereby influencing the earth's climate by making clouds more or less reflective and more or less long-lived. Climatic impacts resulting from interactions between aerosols and clouds have been identified by the IPCC Fourth Assessment Report (AR4; 2007) as being highly uncertain and it is toward the improved representation of these processes in climate models that BNL's efforts are directed.

The parent data set from which the Excel spreadsheet has been derived is archived at the BNL anonymous ftp site:

ftp://ftp.asd.bnl.gov/pub/ASP%20Field%20Programs/2008VOCALS/Processed_Data/PCASP_BPart/

Data are archived as ASCII files.

4.2.1 VOCALS 2008 PCASP-B Aerosol Size Spectrum

Data from the DOE G-1 Research Aircraft Facility operating during the 2008 VAMOS Ocean Cloud Atmosphere Land Study (VOCALS) 2008 based in part at Chacalluta Airport (ARI) north of Arica, Chile were used.

PCASP_BPart—contains detailed size-binned (30 bins, 0.1–3 μm diameter) data obtained from the PCASP (Passive Cavity Aerosol Spectrometer Probe, Unit B). This probe was on the isokinetic inlet in the cabin before 10/29/08. On flight 081029a it was moved to the nose pylon of the plane.

The following 17 data sets were used for the study, the file names giving Flight Designation (yymmdd{flight of day letter}), 081014a_10.txt, 081017a_10.txt, 081018a_10.txt, 081022a_10.txt, 081023a_10.txt, 081025a_10.txt, 081026a_10.txt, 081028a_10.txt, 081029a_10.txt, 081101a_10.txt, 081103a_10.txt, 081104a_10.txt,

081106a_10.txt, 081108a_10.txt, 081110a_10.txt, 081112a_10.txt, 081113a_10.txt. The letter is 'a' for first flight. Maximum data frequency is 10 s^{-1} indicated as '_10' in the file name.

4.3 Analysis and Discussion of Results

The atmospheric suspended particulate size spectrum is closely related to the vertical velocity spectrum (Sect. 1.6). The mean volume radius of suspended aerosol particulates increases with height (or reference level z) in association with decrease in number concentration. At any height (or reference level) z, the fractal fluctuations (of wind, temperature, etc.) carry the signatures of eddy fluctuations of all size scales since the eddy of length scale z encloses smaller scale eddies and at the same time forms part of internal circulations of eddies larger than length scale z (Sect. 1.4). The observed atmospheric suspended particulate size spectrum also exhibits a decrease in number concentration with increase in particulate radius. At any reference level z of measurement the mean volume radius r_{as} will serve to calculate the normalized radius r_{an} for the different radius class intervals as explained below.

The general systems theory for fractal space–time fluctuations in dynamical systems predicts universal mass size spectrum for atmospheric suspended particulates (Sect. 1.6.4). For homogeneous atmospheric suspended particulates, i.e., with the same particulate substance density, the atmospheric suspended particulate mass and radius size spectrum is the same and is given as (Sect. 1.6.4) the normalized aerosol number concentration equal to $\dfrac{1}{N}\dfrac{dN}{d(lnr_{an})}$ versus the normalized aerosol radius r_{an}, where (i) r_{an} is equal to $\dfrac{r_a}{r_{as}}$, r_a being the mean class interval radius and r_{as} the mean volume radius for the total aerosol size spectrum, (ii) N is the total aerosol number concentration and dN is the aerosol number concentration in the aerosol radius class interval dr_a, and (iii) $d(\ln r_{an})$ is equal to $\dfrac{dr_a}{r_a}$ for the aerosol radius class interval r_a to $r_a + dr_a$.

4.3.1 Analysis Results, VOCALS PCASP-B Aerosol Size Spectrum

A total of 17 data sets between 14 October and 13 November 2008 are available for the study. The data used in this study for each of the 17 flights are (i) average and standard deviation for particle number concentration per cc in 29 class intervals ranging from.1 to 3 μm for the particle diameter, (ii) average and standard deviation for total particle number concentration per cc (bins 2–30), and (iii) average and standard deviation for total volume (cc).

Details of data sets used for the study are shown in Figs. 4.1 (a–d) as follows. (i) Fig. 4.1a: lower and upper radius size limits for bin numbers 2–30, (ii) Fig. 4.1b: average and standard deviation for total particle number concentration per cc for the 17 flights, (iii) Fig. 4.1c: average and standard deviation for total volume (cc)

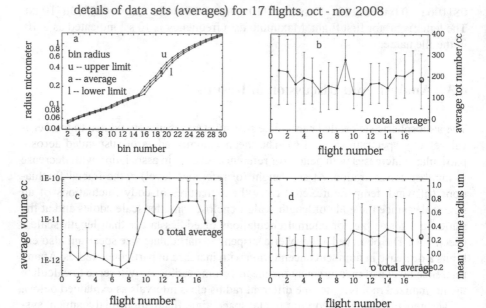

details of data sets (averages) for 17 flights, oct - nov 2008

error bars indicate one standard deviation on either side of mean

Fig. 4.1 **a** Lower and upper radius size limits for bin numbers 2–30. **b** Average and standard deviation for total particle number concentration per cc for the 17 flights. **c** Average and standard deviation for total volume (cc) for the 17 flights (bins 2–30). **d** Average and standard deviation for mean volume radius (μm) for the 17 flights

for the 17 flights (bins 2–30), and (iv) Fig. 4.1d: average and standard deviation for mean volume radius (μm) for the 17 flights. The dispersion (equal to standard deviation/mean) expressed as percentage gives a statistical measure of variability of measured particle number concentration. Computed dispersion (%) values are plotted for the two size ranges (i) less than 1 μm diameter (bins 2–20) and (ii) 1–3 μm diameter (bins 21–30) in Fig. 4.2a, b respectively.

The average total number concentration exhibits a variability of about ± 100 cc^{-1} around a mean value of about 200 cc^{-1} except for the first three flights which show larger variability (Fig. 4.1b). The total volume is one order of magnitude larger for flight numbers 10 onwards compared to earlier flights (Fig. 4.1c) consistent with larger median volume radii for flight numbers 10 onwards (Fig. 4.1d) and exhibit large variability, particularly for size ranges more than 1 μm (Fig. 4.1d).

For particle diameter range less than 1 μm (bins 2–20) the computed dispersion (%) for particle number concentration is within 100 % for bins 2–14 size range and thereafter increases rapidly to a maximum of 500 %. The computed dispersion (%) for particle number concentration for bins 21–30 (1–3 μm diameter) increases steeply from 500 %–5000 % with increase in particle size.

The mid-point diameter of the class interval was used to compute the corresponding value of d(lnr_{an}). The average aerosol size spectra for each of the 17 data sets are plotted on the left-hand side and the total average spectrum for the 17 data sets is plotted on the right-hand side in Fig. 4.3 along with the model-predicted

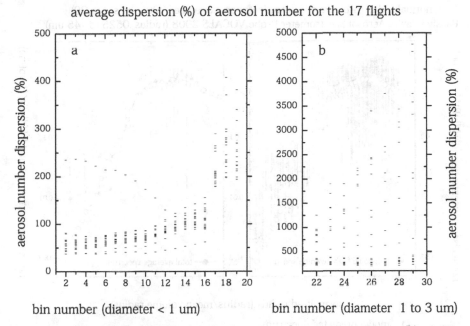

average dispersion (%) of aerosol number for the 17 flights

Fig. 4.2 Computed dispersion (%) values for size range. **a** Less than 1 μm diameter (bins 2–20). **b** 1–3 μm diameter (bins 21–30)

normalised aerosol size spectrum october, november 2008, 17 flights
Passive Cavity Aerosol Spectrometer Probe VOCALS 2008 (.1 - 3 um)

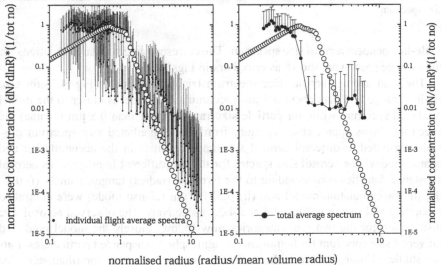

○   model predicted spectrum
error bars indicate one standard deviation on either side of mean

Fig. 4.3 Average aerosol size spectra (bins 2–30) for each of the 17 data sets are plotted on the left-hand side and the total average spectrum for the 17 data sets is plotted on the right along with the model-predicted scale-independent aerosol size spectrum

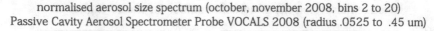

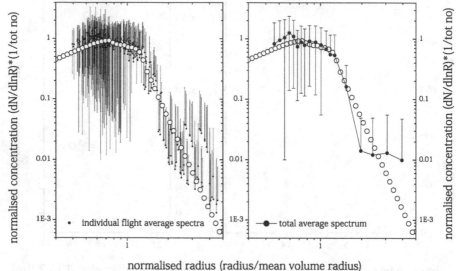

normalised radius (radius/mean volume radius)

○   model predicted spectrum
error bars indicate one standard deviation on either side of mean

Fig. 4.4 Aerosol size spectra for homogeneous aerosol substance density in the accumulation mode corresponding to the size (radius) range 0.1–0.5 μm (bins 2–20). The average aerosol size spectra for each of the 17 data sets are plotted on the left-hand side and the total average spectrum for the 17 data sets is plotted on the right along with the model-predicted scale-independent aerosol size spectrum

scale-independent aerosol size spectrum. The corresponding standard deviations for the average spectra are shown as error bars in Figs. 4.3.

The total average aerosol size spectrum (right-hand side of Fig. 4.3) for size (radius) range less than about 0.5 μm (accumulation mode) is closer to the model predicted spectrum while for particle size range greater than 0.5 μm (coarse) the spectrum shows appreciable departure from model-predicted size spectrum possibly attributed to different aerosol substance densities in the accumulation and coarse modes. The aerosol size spectra for the two different homogeneous aerosol substance densities corresponding to the two size (radius) ranges, namely (i) 0.1–0.5 μm (accumulation mode) and (ii) 0.5–1.5 μm (coarse mode) were computed separately and shown in Figs. 4.4 and 4.5, respectively. The observed aerosol size distribution for the two size categories now follows closely the model predicted universal size spectrum for homogeneous atmospheric suspended particulates. Earlier studies (Husar 2005)have shown that the source for submicron (diameter) size accumulation mode aerosols is different from the larger (greater than 1 μm diameter) coarse mode particles in the atmosphere and therefore may form two different homogeneous aerosol size groups.

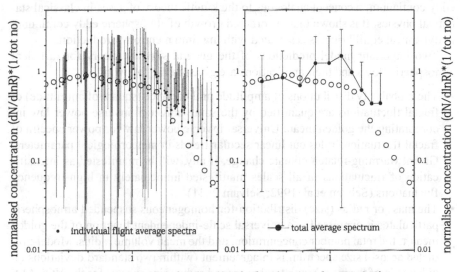

normalised aerosol size spectrum (october, november 2008, bins 21 to 30)
Passive Cavity Aerosol Spectrometer Probe VOCALS 2008 (radius .55 to 1.45 um)

normalised radius (radius/mean volume radius)

○   model predicted spectrum
error bars indicate one standard deviation on either side of mean

Fig. 4.5 Aerosol size spectra for homogeneous aerosol substance density in the coarse mode corresponding to the size (radius) range 0.5–1.5 μm (bins 21–30). The average aerosol size spectra for each of the 17 data sets are plotted on the left-hand side and the total average spectrum for the 17 data sets is plotted on the right along with the model-predicted scale-independent aerosol size spectrum

The amount and longitudinal gradient of aerosol sulfate, and a consideration of the locations of Cu smelters and power plants in Chile, strongly suggest that the submicron aerosol is dominated by anthropogenic emissions (Kleinman et al. 2009).

The source of the uncertainties displayed by the error bars may be due to measurement noise, independent in every size interval, also may be due to different aerosol sources. The model predicts universal spectrum for suspended aerosol mass size distribution(Sect. 1.6.4), based on the concept that the atmospheric eddies hold in suspension the aerosols and thus the mass size spectrum of the atmospheric aerosols is dependent on the vertical velocity fluctuation spectrum of the atmospheric eddies.

Conclusions

The apparently irregular (turbulent) atmospheric flows exhibit self-similar fractal fluctuations associated with inverse power law distribution for power spectra of meteorological parameters on all time scales signifying an eddy continuum underlying the fluctuations. A general systems theory (Selvam 1990) visualizes each large eddy as the envelope (average) of enclosed smaller scale eddies, thereby generating the eddy continuum, a concept analogous to the kinetic theory of gases in classical statistical physics. It is shown that the ordered growth of atmospheric eddy continuum in dynamical equilibrium is associated with maximum entropy production.

Two important model predictions of the general systems theory for turbulent atmospheric flows and their applications are given in the following:

- The probability distributions of amplitude and variance (square of amplitude) of fractal fluctuations are quantified by the same universal inverse power law incorporating the golden mean. Universal inverse power law for power spectra of fractal fluctuations rules out linear secular trends in meteorological parameters. Global warming-related climate change, if any, will be manifested as intensification of fluctuations of all scales manifested immediately in high-frequency fluctuations (Selvam et al. 1992; Selvam 2011)
- The mass or radius (size) distribution for homogeneous suspended atmospheric particulates is expressed as a universal scale-independent function of the golden mean τ, the total number concentration, and the mean volume radius. Model-predicted aerosol size spectrum is in agreement (within two standard deviations on either side of the mean) with total averaged radius size spectra for the VOCALS 2008 PCASP-B data sets. SAFARI 2000 aerosol size distributions reported by Haywood et al. (2003) also show similar shape for the distributions. Specification of cloud droplet size distributions is essential for the calculation of radiation transfer in clouds and cloud-climate interactions, and for remote sensing of cloud properties. The general systems theory model for aerosol size distribution is scale-free and is derived directly from atmospheric eddy dynamical concepts. At present empirical models such as the log normal distribution with arbitrary constants for the size distribution of atmospheric suspended particulates is used for quantitative estimation of the earth-atmosphere radiation budget related to climate warming/cooling trends (Sect. 2.1). The universal aerosol size spectrum presented in this chapter may be computed for any location with two measured parameters, namely, the mean volume radius and the total number concentration and may be incorporated in climate models for computation of radiation budget of earth-atmosphere system.

References

Haywood J, Francis P, Dubovik O, Glew M, Holben B (2003) Comparison of aerosol size distributions, radiative properties, and optical depths determined by aircraft observations. J Geophys Res 108(D13) 8471:SAF 7–1 to 12

Husar RB (2005) The emergence of the bimodal distribution concept. In: Sem GJ, Boulaud D, Brimblecombe P, Ensor ES, Gentry JW, Marijnissen JCM, Preining O (eds) History reviews of aerosol science. American Association for Aerosol Research, Mt. Laurel

IPCC Fourth Assessment Report (AR4) (2007) Climate change 2007. The physical science basis. In: Solomon S, Qin D, Manning M, Chen Z, Marquis M, Averyt KB, Tignor M, Miller HL (eds), Contribution of Working Group I to the Fourth Assessment Report of the Intergovernmental Panel on Climate Change. Cambridge University Press, Cambridge

Joshi RR, Selvam AM (1999) Identification of self-organised criticality in atmospheric low frequency variability. Fractals 7(4):421–425

Kleinman LI, Springston SR, Daum PH, Lee Y-N, Sedlacek AJ, Senum G, Wang J (2009) Pre-cloud aerosol, cloud droplet concentration, and cloud condensation nuclei from the VAMOS Ocean-Cloud-Atmosphere L and Study (VOCALS) Field Campaign, First Quarter 2010, ASR Program Metric Report pp. 1–11, December 2009, Research sponsored by the U.S. Department of Energy Office of Science, Office of Biological and Environmental Research

Selvam AM (1990) Deterministic chaos, fractals and quantumlike mechanics in atmospheric flows. Can J Phys 68:831–841. http://xxx.lanl.gov/html/physics/0010046

Selvam AM (2005) A general systems theory for chaos, quantum mechanics and gravity for dynamical systems of all space-time scales. Electromagnetic Phenomena 5 No.2(15):160–176 http://arxiv.org/pdf/physics/0503028; http://www.emph.com.ua/15/selvam.htm

Selvam AM (2007) Chaotic climate dynamics. Luniver, UK

Selvam AM (2009) Fractal fluctuations and statistical normal distribution. Fractals 17(3):333–349. http://arxiv.org/pdf/0805.3426

Selvam AM (2011) Signatures of universal characteristics of fractal fluctuations in global mean monthly temperature anomalies. J Syst Sci and Complexity 24(1):14–38 http://arxiv.org/abs/0808.2388

Selvam AM (2012) Universal spectrum for atmospheric aerosol size distribution: comparison with pcasp-b observations of vocals 2008. Nonlinear Dynamics and Systems Theory 12(4):397–434. http://arxiv.org/abs/1105.0172

Selvam AM, Fadnavis S (1998) Signatures of a universal spectrum for atmospheric inter-annual variability in some disparate climatic regimes. Mcteor Atmos Phys 66:87–112. http://xxx.lanl.gov/abs/chao-dyn/9805028

Selvam AM, Joshi RR (1995) Universal spectrum for interannual variability in COADS global air and sea surface temperatures. Int J Climatol 15:613–623

Selvam AM, Pethkar JS, Kulkarni (1992) M.K Signatures of a universal spectrum for atmospheric interannual variability in rainfall time series over the Indian Region. Int J Climatol 12:137–152

Selvam AM, Pethkar JS, Kulkarni MK, Vijayakumar R (1996) 1996: Signatures of a universal spectrum for atmospheric interannual variability in COADS surface pressure time series. Int J Climatol 16:393–404

Chapter 5
Universal Spectrum for Atmospheric Suspended Particulates: Comparison with Observations: Data Set III

Abstract Atmospheric flows exhibit fractal fluctuations and inverse power law for power spectra indicates an eddy continuum structure for the self-similar fluctuations. A general systems theory for aerosol size distribution based on fractal fluctuations is proposed. The model predicts universal (scale-free) inverse power law for fractal fluctuations expressed in terms of the golden mean. Atmospheric particulates are held in suspension in the fractal fluctuations of vertical wind velocity. The mass or radius (size) distribution for homogeneous suspended atmospheric particulates is expressed as a universal scale-independent function of the golden mean, the total number concentration, and the mean volume radius. Model-predicted spectrum is compared with the total averaged radius size spectra for the AERONET (aerosol inversions) stations Davos and Mauna Loa for 2010 and Izana for 2009. There is close agreement between the model predicted and the observed aerosol spectra. The proposed model for universal aerosol size spectrum will have applications in computations of the radiation balance of earth–atmosphere system in climate models.

Keywords Universal spectrum for aerosol size distribution · General systems theory for atmospheric aerosol size spectrum · AERONET (aerosol inversions) stations Davos, Mauna Loa, and Izana · Nonlinear dynamics and chaos, radiation balance of earthatmosphere system in climate models

5.1 Introduction

Atmospheric aerosol mass size spectrum is expressed in terms of the universal eddy energy spectrumand is a function of the mean volume radius and total number concentration. Model predictions are in agreement with observed daily mean Aeronet aerosol size spectrum at Davos and Mauna Loa for 2010 and at Izana for 2009 (Selvam 2013).

5.2   Data

Daily mean volume particle size distribution $dV(r)/d\ln r$ ($\mu m^3/\mu m^2$) retrieved in 22 logarithmically equidistant bins in the range of sizes $0.05 \, \mu m \le r \le 15 \, \mu m$ for Davos (Switzerland) and Mauna Loa (Hawaii) for the year 2010 and Izana (Spain) for the year 2009 were obtained from AERONET aerosol robotic network (http://aeronet.gsfc.nasa.gov/new_web/data.html). For Izana (Spain), the complete data set for the 12-month period of 2009 was available and hence it was used. Daily average data are calculated from all points for each day when three (3) or more points are available. Data sets for a total of 54, 180, and 133 days, respectively, were available for the three stations Davos, Mauna Loa, and Izana. The formulas for calculating standard parameters of the particle size distribution (http://aeronet.gsfc.nasa.gov/new_web/Documents/Inversion_products_V2.pdf) are given below.

AERONET retrieves the aerosol size distribution of the particle volume $dV(r)/d\ln r$. It relates to the distribution of particle number as follows:

$$\frac{dV(r)}{d\ln(r)} = V(r)\frac{dN(r)}{d\ln(r)} = \frac{4}{3}\pi r^3 \frac{dN(r)}{d\ln(r)}.$$

Volume concentration ($\mu m^3/\mu m^2$):

$$C_v = \int_{r_{\min}}^{r_{\max}} \frac{dV(r)}{d\ln(r)}.$$

The details of computations in the present study are as follows:

1. The mid-point (center of size bin) radius r was used to calculate the distribution of particle number $dN(r)/d\ln(r)$ from the volume concentration $dV(r)/d\ln(r)$ for fine (f) and course (c) aerosol modes.
2. The radius range for fine (f) mode is $0.05 < r \le 0.6 \, \mu m$ and the radius range for the coarse (c) mode is $0.6 < r \le 15 \, \mu m$.

The total number concentrations $N\left(\sum dN(r)\right)$ for fine (f) and coarse (c) modes were calculated from $dN(r)/d\ln(r)$ since $d\ln(r)$ is a constant equal to $d(r)/r$ for the retrieved size spectrum with logarithmically equidistant bins. The constant $d\ln(r)$ computed as equal to $(r_2 - r_1)/r_1$ from the known values of mid-point (center of size bin) radii r is equal to 0.31207.

5.3   Analysis and Results

The atmospheric suspended particulate size spectrum is closely related to the vertical velocity spectrum (Sect. 1.6). The mean volume radius of suspended aerosol particulates increases with height (or reference level z) in association with decrease in number concentration. At any height (or reference level) z, the fractal fluctuations (of wind, temperature, etc.) carry the signatures of eddy fluctuations of all size scales since the eddy of length scale z encloses smaller scale eddies and at the same time forms part of internal circulations of eddies larger than length scale z (Sect. 1.4). The observed atmospheric suspended particulate size spectrum also exhibits a decrease in number concentration with increase in particulate radius. At any reference level z of measurement the mean volume radius r_{as} will serve to calculate the normalized radius r_{an} for the different radius class intervals as explained below.

The general systems theory for fractal space–time fluctuations in dynamical systems predicts universal mass size spectrum for atmospheric suspended particulates (Sect. 1.6.4). For homogeneous atmospheric suspended particulates, i.e., with the same particulate substance density, the atmospheric suspended particulate mass and radius size spectrum is the same and is given as the normalized aerosol number concentration equal to $\dfrac{1}{N}\dfrac{dN}{d(lnr_{an})}$ versus the normalized aerosol radius r_{an}, where (i) r_{an} is equal to $\dfrac{r_a}{r_{as}}$, r_a being the mean class interval radius and r_{as} the mean volume radius for the total aerosol size spectrum, (ii) N is the total aerosol number concentration and dN is the aerosol number concentration in the aerosol radius class interval dr_a, and (iii) $d(\ln r_{an})$ is equal to $\dfrac{dr_a}{r_a}$ for the aerosol radius class interval r_a to $r_a + dr_a$.

The average normalized aerosol size spectra for fine (f) mode for Davos, Mauna Loa, and Izana with 54, 180, and 133 daily mean data sets, respectively, are shown in Fig. 5.1 along with the model predicted universal normalized aerosol size spectrum. The corresponding aerosol size spectra for coarse (c) mode are given in Fig. 5.2. The observed aerosol size spectra are in close agreement with model-predicted universal spectrum for suspended particulates in the turbulent atmospheric flows. The total average mean volume radius and total number concentration for the three stations for the period of study are given in Fig. 5.3. The mean volume radius and total number concentration are minimum for Mauna Loa (Hawaii) for both fine and coarse aerosol modes. Coarse mode particulate number concentration is a maximum for Izana (Spain).

Discussion and Conclusions

Atmospheric flows exhibit self-similar fractal fluctuations on all space–time scales. Fractal fluctuations are ubiquitous to dynamical systems in nature such as fluid flows, heart beat patterns, population growth, etc. Power spectra of fractal fluctuations exhibit inverse power law form indicating long-range correlations, identified as self-organised criticality. Identification and quantification of the exact physical laws underlying the observed self-organised criticality will help predict the future evolution of dynamical systems such as weather patterns. A general systems theory which satisfies the maximum entropy principle of classical statistical physics recently proposed by the author enables formulation of precise quantitative relations

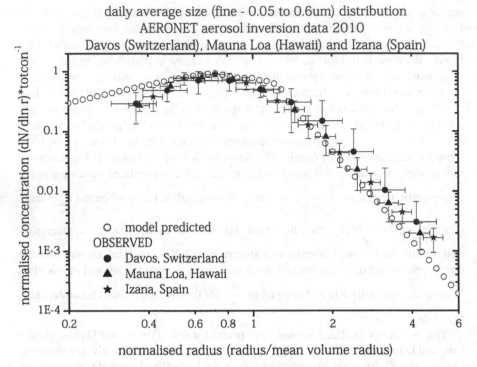

Fig. 5.1 Mean daily average aerosol size distribution, fine mode (0.05 < radius ≤ 0.6 μm) for (*1*) Davos for 2010, (*2*) Mauna Loa for 2010, and (*3*) Izana for 2009

for the observed universal characteristics of fractal fluctuations in turbulent atmospheric flows. The model predictions are as follows. (i) The apparently chaotic (unpredictable) fluctuations can be resolved into a nested continuum of vortex roll circulations tracing the space filling quasiperiodic Penrose tiling pattern with an overall logarithmic spiral trajectory. (ii) The amplitude and also the variance (square of the amplitude) of fractal fluctuations are quantified by the same statistical probability distribution function incorporating the golden mean τ and exhibits scale-free universal inverse power law characteristics. Therefore, fractal fluctuations are signatures of quantum-like chaos since square of the eddy amplitudes, i.e., variances, represent the probability densities (of amplitudes), a property exhibited by the subatomic dynamics of quantum systems such as electron or photon. (iii) Atmospheric particulates are held in suspension in the fractal fluctuations of vertical wind velocity. The mass or radius (size) distribution for homogeneous suspended atmospheric particulates is expressed as a universal scale-independent function of the golden mean τ, the total number concentration, and the mean volume radius. The universal aerosol size spectrum will have applications in computation of radiation balance of earth–atmosphere system in climate models.

Model predictions are in agreement separately, with the fine (0.1–0.6 μm) and the coarse (0.6–15 μm) mode AERONET aerosol inversion data sets (daily averages)

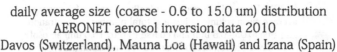

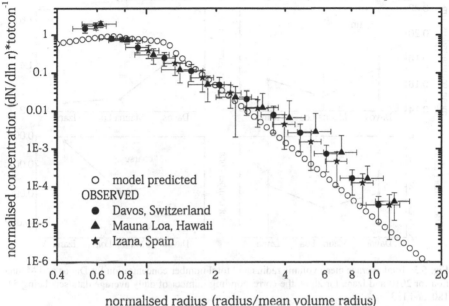

Fig. 5.2 Mean daily average aerosol size distribution, coarse mode ($0.6 < \text{radius} \leq 15.0 \ \mu\text{m}$) for Davos and Mauna Loa for 2010 and Izana for 2009

for Davos and Mauna Loa for 2010 and Izana for 2009. The results indicate two different homogeneous aerosol substance densities corresponding to the two size (radius) ranges, namely (i) 0.1–0.6 μm (fine mode) and (ii) 0.6–15 μm (coarse mode). Earlier studies (Husar 2005) have shown that the source for submicron (diameter) size accumulation mode aerosols is different from the larger (greater than 1 μm diameter) coarse mode particles in the atmosphere and therefore may form two different homogeneous aerosol size groups.

The source of the uncertainties displayed by the error bars (Figs. 5.1 to 5.3) may be due to measurement noise, independent in every size interval, also may be due to different aerosol sources. The model predicts universal spectrum for suspended aerosol mass size distribution (Sect. 1.6.4), based on the concept that the atmospheric eddies hold in suspension the aerosols and thus the mass size spectrum of the atmospheric aerosols is dependent on the vertical velocity fluctuation spectrum of the atmospheric eddies.

At present empirical models such as the log normal distribution with arbitrary constants for the size distribution of atmospheric suspended particulates are used for quantitative estimation of earth–atmosphere radiation budget related to climate warming/cooling trends. The general systems theory model for aerosol size distribution is scale-free and is derived directly from atmospheric eddy dynamical concepts.

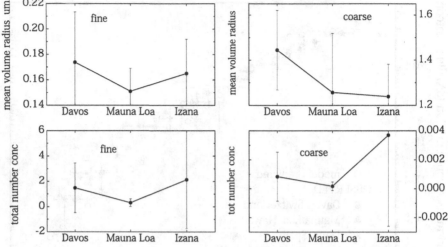

Fig. 5.3 Total average mean volume radius and total number concentration for Davos and Mauna Loa for 2010 and Izana for 2009, the corresponding number of daily average data sets being 54, 180, and 133

The universal aerosol size spectrum, a function of the total aerosol number concentration and the mean volume radius alone, will have applications in computation of radiation balance of earth–atmosphere system in climate models.

Acknowledgements The author is grateful to Dr. Christoph_Wehrli, Dr. Brent. N. Holben, and Dr. Philippe Goloub, the respective principal investigators of AERONET sites Davos, Mauna Loa, and Izana and their team members for free and generous distribution of aerosol data to the scientific community and their efforts in establishing and maintaining the AERONET sites.

References

Husar RB (2005) The emergence of the bimodal distribution concept. In: Sem GJ, Boulaud D, Brimblecombe P, Ensor ES, Gentry JW, Marijnissen JCM, Preining O (eds) History reviews of aerosol science. American Association for Aerosol Research, Mount Laurel
Selvam AM (2013) Scale-free Universal Spectrum for Atmospheric Aerosol Size Distribution for Davos, Mauna Loa and Izana. Int J Bifurcation Chaos 23, 1350028. http://arxiv.org/pdf/1111.3132

Chapter 6
Universal Spectrum for Atmospheric Suspended Particulates: Comparison with Observations, Data Set IV

Abstract Atmospheric flows exhibit self-similar fractal space-time fluctuations manifested as the fractal geometry to global cloud cover pattern and inverse power law form for power spectra of meteorological parameters such as windspeed, temperature, rainfall, etc. Inverse power law form for power spectra indicates long-range space-time correlations or non-local connections and is a signature of self-organiszed criticality generic to dynamical systems in nature. The general systems theory discussed in Chaps. 1 and 2 predicts the observed self-organiszed criticality as a signature of quantum-like chaos in dynamical systems. The model predictions are as follows. (i) The fractal fluctuations can be resolved into an overall logarithmic spiral trajectory with the quasiperiodic Penrose tiling pattern for the internal structure. (ii) The probability distribution represents the power (variance) spectrum for fractal fluctuations and follows universal inverse power law form incorporating the *golden mean*. Such a result that the additive amplitudes of eddies when squared represent probability distribution is observed in the subatomic dynamics of quantum systems such as the electron or photon. Therefore, the irregular or unpredictable fractal fluctuations exhibit quantum-like chaos. (iii) Atmospheric aerosols are held in suspension by the vertical velocity distribution (spectrum). The atmospheric aerosol size spectrum is derived in terms of the universal inverse power law characterizing atmospheric eddy energy spectrum. Model- predicted spectrum is in agreement with the following two experimentally determined atmospheric aerosol data sets, (i) SAFARI 2000 CV-580 Aerosol Data, Dry Season 2000 (CARG), (ii) World Data Centre Aerosols data sets for the three stations Ny Ålesund, Pallas, and Hohenpeissenberg.

Keywords Universal atmospheric aerosol size spectrum · SAFARI 2000 aerosol size spectra · World data center aerosol size spectra · Fractal fluctuations in atmospheric flows · Chaos and nonlinear dynamics

© The Author(s) 2015
89
A. M. Selvam, *Rain Formation in Warm Clouds,* SpringerBriefs in Meteorology,
DOI 10.1007/978-3-319-13269-3_6

6.1 Introduction

Atmospheric aerosols are held in suspension by the vertical velocity distribution (spectrum). The normalized atmospheric aerosol size spectrum is derived in terms of the universal inverse power law characterizing atmospheric eddy energy spectrum. Model-predicted spectrum is in agreement with the following two experimentally determined atmospheric aerosol data sets, (i) SAFARI 2000 CV-580 Aerosol Data, Dry Season 2000 (CARG), (ii) World Data Centre Aerosols data sets for the three stations Ny Ålesund, Pallas, and Hohenpeissenberg (Selvam 2011).

6.2 Data

The following two data sets were used for comparison of observed with model-predicted aerosol size spectrum.

6.2.1 Data Set I

SAFARI 2000 CV-580 Aerosol Data, Dry Season 2000 (CARG). The Cloud and Aerosol Research Group (CARG) of the University of Washington participated in the SAFARI-2000 Dry Season Aircraft campaign with their Convair-580 research aircraft. This campaign covered five countries in southern Africa from August 10 to September 18, 2000. Various types of measurements were obtained on the 31 research flights of the Convair-580 to study their relationships to simultaneous measurements from satellites (particularly Terra), other research aircraft, and SAFARI-2000 ground-based measurements and activities. The UW technical report for the SAFARI 2000 project (http://daac.ornl.gov/data/safari2k/atmospheric/CV-580/comp/SAFARI-MASTER.pdf) gives a complete detailed guide to the extensive measurements obtained aboard the UW Convair-580 aircraft in support of SAFARI 2000 (Hobbs 2004).

Aerosol Data (aer) instrument details are given in Table 6.1

Data sets satisfying the following four conditions were chosen for analysis. (i) Flights where both *pcasp* and *tsi3320* are available. (ii) There are no unavailable (−999.99) data in any of the size intervals. (iii) The number of class intervals with zero aerosol number concentration does not exceed three within the first five class intervals for *pcasp* for data inclusion for both *pcasp* and *tsi3320*. (iv) The standard deviation of aerosol number concentrations in the class intervals of the size spectrum is less than 0.8 in order to exclude any abnormally large fluctuations, particularly in the tail end of the spectrum. Aerosol data sets for 21 aircraft flights were used for the study and details of available number of aerosol size spectra for each flight are given in Table 6.2. The available number of data values for each of the 15 size class intervals for *pcasp* and 52 size class intervals for *tsi3320* instrumentation systems are shown in Fig. 6.1.

Table 6.1 Aerosol Data (aer) instrument details. (ftp://ftp.daac.ornl.gov/data/safari2k/atmospheric/CV-580/comp/CV-580.pdf)

Instrument I	
Name	pcaspdnc
Definition	Particle concentration between 0.1 and 3.0 μm in 15 channels (corrected per micron)
Units	#/cc
Instrument	Particle measuring systems model PCASP-100X
Processing	Calculated from raw counts and sample time
Notes	Channel limits are: 0.10, 0.12, 0.14, 0.17, 0.20, 0.25, 0.30, 0.40, 0.50, 0.70, 0.90, 1.20, 1.50, 2.00, 2.50, 3.00 μm
Instrument II	
Name	tsidnc
Definition	Particle concentration between 0.5 and 20 μm in 52 channels (corrected per micron)
Units	#/cc
Instrument	TSI Model 3320 APS
Processing	Calculated from raw counts and sample time
Notes	There are no data from this instrument on UW flights 1815, 1818, 1820, and 1830. Channel limits are: 0.487, 0.523, 0.562, 0.604, 0.649, 0.698, 0.750, 0.806, 0.866, 0.931, 1.000, 1.075, 1.155, 1.241, 1.334, 1.433, 1.540, 1.655, 1.778, 1.911, 2.054, 2.207, 2.371, 2.548, 2.738, 2.943, 3.162, 3.398, 3.652, 3.924, 4.217, 4.532, 4.870, 5.233, 5.623, 6.043, 6.494, 6.978, 7.499, 8.058, 8.660, 9.306, 10.000, 10.746, 11.548, 12.409, 13.335, 14.330, 15.399, 16.548, 17.783, 19.110, 20.535 μm

6.2.2 Data Set II

Aerosol size distributions for three land stations (Ny Ålesund, Pallas, and Hohenpeissenberg) were obtained from World Data Centre for Aerosols (http://wdca.jrc.it/data/parameters/data_size.html) at The aerosol size distribution data archive and details of these data sets are given in Table 6.3. (Archive: ftp://ftp-ccu.jrc.it/pub/WDCA/NARSTO_archive/2.301/parameters/Size_Distribution). The annual means and corresponding standard deviations of aerosol number concentrations for the aerosol size class intervals were used for the study.

6.3 Analysis

The aerosol size spectrum is given as (Sect. 1.6.4) the normalized aerosol number concentration equal to $\dfrac{1}{N}\dfrac{dN}{d(lnr_{an})}$ versus the normalized aerosol radius $r_{an,}$ where

(i) r_{an} is equal to $\dfrac{r_a}{r_{as}}$, r_a being the mean class interval radius and r_{as} the mean

Table 6.2 Safari 2000 CV-580 aerosol size spectra, dry season 2000 (CARG)

Details of available number of aerosol size spectra

No	Flight	Year	Month	Day	Pcasp—15 class intervals	tsi3320—52 class intervals
					No of spectra	No of spectra
1	1810	2000	08	10	12,192	10,364
2	1812	2000	08	14	9200	8165
3	1813	2000	08	14	2615	1481
4	1814	2000	08	15	11,615	4359
5	1816	2000	08	18	7738	6840
6	1819	2000	08	20	14,801	13,740
7	1821	2000	08	23	9865	6605
8	1823	2000	08	29	9016	7486
9	1824	2000	08	29	9329	9201
10	1825	2000	08	31	18,709	13,688
11	1826	2000	09	01	4409	4408
12	1828	2000	09	01	7004	5993
13	1829	2000	09	02	20,153	19,477
14	1831	2000	09	05	18,806	18,756
15	1832	2000	09	06	13,275	12,967
16	1833	2000	09	06	7449	7380
17	1834	2000	09	07	14,376	4312
18	1835	2000	09	10	14,727	13,862
19	1836	2000	09	11	11,988	9039
20	1838	2000	09	14	11,529	3968
21	1839	2000	09	16	16,241	7670
Total					245,037	189,761

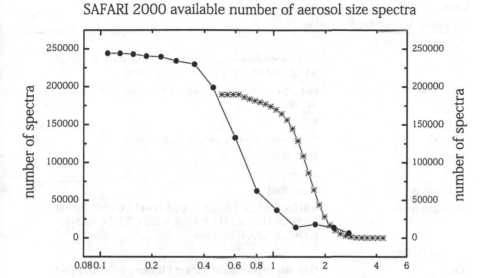

Fig. 6.1 Available number of data values for each of the 15 size class intervals for *pcasp* and 52 size class intervals for *tsi3320* instrumentation systems

volume radius for the aerosol size spectrum (reference level). (ii) N is the total aerosol number concentration and dN is the aerosol number concentration in the aerosol radius class interval dr_a. (iii) $d(\ln r_{an})$ is equal to $\dfrac{dr_a}{r_a}$ for the aerosol radius class interval r_a to $r_a + dr_a$.

6.3.1   Data I: SAFARI 2000 CV-580 Aerosol Size Spectra, Dry Season 2000 (CARG)

The mean and standard deviation of normalized aerosol size spectrum were computed for 245,037 and 189,761 individual spectra (Table 6.1), respectively, for *pcasp* and *tsi3320* aerosol measurement instrument systems and shown in Fig. 6.2 along with the model-predicted universal normalized aerosol size spectrum.

6.3.2   Data II: World Data Centre for Aerosols

The annual mean normalized aerosol size spectra with associated standard deviation were computed for the three stations (Table 6.3) Ny Ålesund, Pallas, and Hohenpeissenberg for each year and shown in Fig. 6.3.

Table 6.3 World data centre aerosols details

Parameter: number size distribution

Station: Ny Ålesund

Data	QAC narsto permanent archive file name WDCA_GAWA_NYAZ_SIZEN_2001/2002/2003/2004.csv
Definition	Particle concentration between 0.017.8 and 0.7079 μm in 16 size classes
Units	#/cc
Method	Differential mobility particle sizer (DMA)
Start	2000/01/01
End	ongoing
Data period	2001–2004
Size-class limits	0.017.8, 0.022.4, 0.028.2, 0.0355, 0.0447, 0.0562, 0.0708, 0.0891, 0.1122, 0.1413, 0.1778, 0.2239, 0.2818, 0.3548, 0.4467, 0.5623, 0.7079 μm

Station: Pallas

Data	QAC narsto permanent archive file name WDCA_GAWA_PAL_SIZEN_2001/2002/2003/2004.csv
Definition	Particle concentration between 0.0069 and 0.53 μm in 30 size classes
Units	#/cc
Method	Differential mobility particle sizer (DMA)
Start	2000/01/01
End	Ongoing
Data period	2001–2004
Size-class limits	0.0069, 0.0079, 0.0092, 0.0106, 0.0123, 0.0142, 0.0164, 0.019, 0.022, 0.0254, 0.0294, 0.034, 0.0393, 0.0454, 0.0525, 0.0607, 0.0703, 0.0811, 0.0937, 0.1084, 0.1252, 0.1446, 0.1671, 0.1931, 0.223, 0.2573, 0.2974, 0.3436, 0.3966, 0.4583, 0.53 μm

Station: Hohenpeissenberg

Data	QAC narsto permanent archive file name WDCA_GAWA_HOP_SIZE_2001/2002/2003/2004/2005.csv
Definition	Particle concentration between 0.1 and 6.75 μm in 15 size classes
Units	#/cc
Method	Optical particle counter (OPC)
Start	1996/10/01
End	Ongoing
Data period	2001–2005
Size-class limits	0.1, 0.12, 0.15, 0.2, 0.25, 0.35, 0.45, 0.6, 0.75, 1, 1.5, 2, 3, 4.5, 6, 7.5 μm

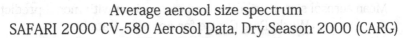

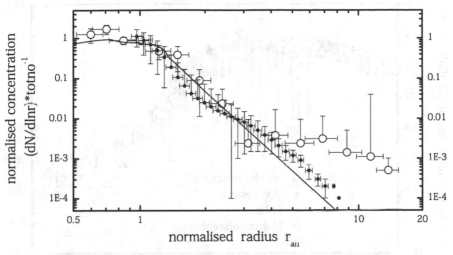

spectra with sdev < 0.8 for total number concentrations
   ○   pcasp concentration corrected per micron 15 class intervals
   •   tsi3320 concentration corrected per micron (tsidnc) 52 class intervals
   ——— model spectrum

Fig. 6.2 Average aerosol size spectrum for SAFARI 2000 CV-580 aerosol size spectra and comparison with model prediction. *Error bars* indicate one standard deviation on either side of the mean

Conclusion

There is a close agreement between the model-predicted and the observed aerosol size distributions for the two aerosol data sets (SAFARI 2000 and World Data Center) used in the study. SAFARI 2000 aerosol size distributions reported by Haywood et al. (2003) also show similar shape for the distributions.

The distribution of atmospheric aerosols is not only determined by turbulence but also by dry and wet chemistry, sedimentation, gas to particle conversion, coagulation, (fractal) variability at the surface, among others. However, at any instant, the mass (and therefore the radius for homogeneous aerosols) size distribution of atmospheric suspensions (aerosols) is directly related to the wind velocity (eddy energy) spectrum, which is shown to be universal (scale independent). The source for aerosols in the fine mode (less than 1 μm) and coarse mode (greater than 1 μm) are different and may account for the observed good fit of the observed radius size spectrum mostly for the fine aerosol mode only.

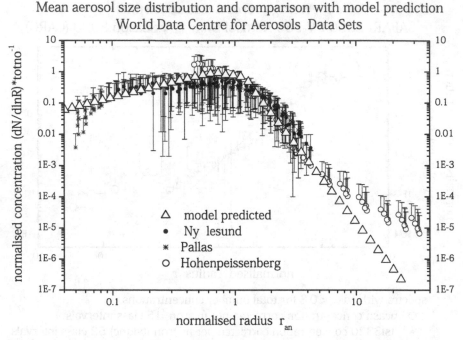

Fig. 6.3 Mean aerosol size spectrum for World data center for aerosols data sets and comparison with model prediction. *Error bars* indicate one standard deviation on either side of the mean

The general systems theory relating to the dynamics of the atmospheric eddy systems proposed in the present chapter can be extended to other planetary, solar, and stellar atmospheres.

Acknowledgment The author is thankful to her husband Dr. A. S. R. Murty for encouragement.

References

Haywood J, Francis P, Dubovik O, Glew M, Holben B (2003) Comparison of aerosol size distributions, radiative properties, and optical depths determined by aircraft observations. J Geophys Res 108 (D13) 8471, SAF 7–1 to 12

Hobbs PV (2004) SAFARI 2000 CV-580 Aerosol and Cloud Data, Dry Season 2000 (CARG). Data set. Available online [http://www.daac.ornl.gov] from Oak Ridge National Laboratory Distributed Active Archive Center, Oak Ridge, Tennessee, U.S.A. doi:10.3334/ORNLDAAC /710

Selvam AM (2011) A general systems theory for atmospheric flows and atmospheric aerosol size distribution. Chaotic Modeling and Simulation (CHAOS 2011 Conference Proceedings) 461–468 retrieved 28 December 2014. http://arxiv.org/ftp/arxiv/papers/0908/0908.2321.pdf

Index

© The Author(s) 2015
A. M. Selvam, *Rain Formation in Warm Clouds*, SpringerBriefs in Meteorology,
DOI 10.1007/978-3-319-13269-3